Prehistory Papers

Cross-disciplinary Studies
Into our Past

Paul Dunbavin

Copyright © 2020 Paul Dunbavin

All rights reserved.

ISBN: 978-0-9525029-4-4

The right of Paul Dunbavin to be identified as the author of this work has been asserted herein in accordance with the Copyright, Designs and Patents Act 1988.

Whilst every effort has been made to identify and acknowledge the source of all quotations and illustrations it may be the case that some have been omitted and we apologise for these instances.

All rights reserved. This book is sold subject to the condition that it shall not, by way of trade or otherwise, be lent, resold, hired out or otherwise circulated without the publisher's prior consent in any form of binding or cover other than that in which it is published and without a similar condition including this condition being imposed on the subsequent purchaser.

British Library Cataloguing in Publication Data
A catalogue record for this book is available from the British Library

All rights reserved.

© Paul Dunbavin and Third Millennium Publishing

ISBN: 978-0-9525029-4-4

BY THE SAME AUTHOR

Atlantis of the West
Picts & Ancient Britons
Under Ancient Skies
Towers of Atlantis

CONTENTS

PREFACE

This book provides a permanent repository for the various articles or papers originally published on the author's website; together with some older articles published, (or unpublished) elsewhere. The articles expand or update the cross-disciplinary research in the author's earlier books

It is in the nature of cross-disciplinary research that it may fail to find a home in a specialist academic journal; perhaps because it would stray outside their normal subject matter, but more likely, because it may refer to evidence that their specialist referee fails to grasp or see the relevance. Such broader enquiry may also challenge long-accepted theories or lead to novel conclusions, which a specialist would not accept.

The articles cover various subject areas to shine a new light on events in the recent human past, including, but not limited-to: astronomy, geophysics, geology, archaeology, mythology, folklore, calendars, climate history, sea-level changes, ancient history, DNA, linguistics and others. There are *no constraints* on where useful evidence about the past may lie; to fill the cracks between the academic disciplines. There is not even a suitable BIC classification under which to categorize a book such as this one!

The primary focus of the author is to examine prehistory, mythology and legends, for which there may be no 'hard' evidence. Often within legends and the snippets of information left by ancient historians, we may find details that are not strictly needed to tell the story, or which contain precise numbers and descriptions where such would not be necessary to give valid background for a fiction. These are the 'mythological fossils' that can be taken-out and separately analysed with the latest science. Quite often the primary source may be an ancient text, for which the modern translator cannot be expected to understand the relevance outside their own field; yet they will disparage any intervention by an outsider who dares to trespass.

Another barrier to progress may be impenetrable specialist jargon and mathematics. When writing for a general audience, or for the benefit of a different specialist, it is often necessary to simplify expert terms to hold the interest of the reader. Again, academics may view this as an indication that the author does not fully understand their subject, while they in turn may have little grasp of evidence drawn from disciplines other than their own. It is not feasible to read every specialist paper or book in every relevant subject; required is a threshold of knowledge in each discipline that is sufficient for the task.

The articles themselves are stand-alone and may differ in style; some were intended as pure interactive web-pages, which give best value online by following the hyperlinks; others were written as short articles for a journal; still others have more references approaching that of formal academic papers. Other than formatting, the articles are presented in their original form. Regardless of style, they are offered in a 'top-down' sequence, with those of mainly astronomical and geophysical content coming before earth-based subjects such as climate, sea-level and archaeology, moving on to those based on human prehistory, with an analysis of myths and legends; and the author's particular interest: *Atlantis*. All the articles were intended to propose original conclusions based on a summary of the evidence analysed within; and inevitably there will be some repetition when they are brought together in the same volume.

It is to be expected that some of the author's conclusions or theories will be mistaken or may not stand the test of time. This should not deter any other researcher from similarly attempting to fill the gaps between the arbitrary boundaries of academic disciplines. If you think that you could do a better job: publish!

1   Dangers to Earth from Ancient Supernovas

This article is less about supernovas and star-mergers themselves, rather the possibility of matter ejected from them impacting the Earth. For the benefit of the general reader scientific terms are kept to a minimum and explained where necessary, or may be pursued in the suggested links. Astrophysical research papers are not usually written for the benefit of the general reader or even for the cross-disciplinary researcher, but for their specialist colleagues. It should not therefore be assumed that the present author fails to understand the terminology and equations in the cited sources; but neither should it be assumed that he understands all of them!

In proposing a theory that the Earth's axis has changed in recent prehistory – be it either a pole-shift or a more extensive change to the obliquity or length-of-day – requires a mechanism to explain how it could happen; the question cannot be ducked. Even a giant asteroid impact like that which caused the extinction of the dinosaurs would not possess enough energy to significantly disturb the Earth's axis and rotation. This has long been sufficient cause for geologists and physicists to dismiss the entire subject of pole shifts and axis-tilts as pseudo-science. The historiography of this subject was discussed in my book *Under Ancient Skies*. It is true; there have been many naïve pole-shift theories which do not acknowledge the vast energy that would be required.

Yet there are clues that something did disturb the Earth's rotation at the close of the ice-age and also again in the mid-

Holocene (see: Raised Beaches and Submerged Forests - Curious Anomalies) To propose a theory of changes to the axis requires an external force that could significantly alter the Earth's angular momentum yet without leaving an obvious impact scar; and without causing a mass extinction event. As discussed in *Under Ancient Skies* there are three principal candidates that might offer an answer to this conundrum:

1) Small-fast meteors ejected by ancient supernovae, stellar mergers and active galaxies.
2) Gravity waves from mergers of compact stars in the close solar neighbourhood
3) As yet unknown physics: dark-matter; gravity leaking from parallel universes, etc.

Here we shall discuss only the first possibility. Since the 1990s there has been much discussion of the collision dangers from near-earth asteroids that we could detect. These might destroy cities and create dust-veils, but would not possess enough energy to significantly change the axis of rotation. Comets could strike with higher velocities up to 72 km/sec for a head-on collision — but while calamitous for all life, even this would not be enough to make the world wobble on its axis. Fast hyperbolic asteroids like *Oumuamua* that may have been ejected from other solar systems may travel even faster, but they still do not possess enough energy. Kinetic energy increases with mass, but with the *square* of the velocity ($E=1/2mv^2$) and so the real danger comes from small fast objects potentially ejected by ancient supernovas. To experiment with the scale of this you may like to try-out this useful calculator tool:

https://www.calculatorsoup.com/calculators/physics/kinetic.php

A little experimentation will show that the velocity of the comet would need to be travelling *at a significant fraction of the speed of light* to approach the kinetic energy of the Earth's rotation, calculated at 2.138×10^{29} joules. Could a supernova or a stellar-merger send objects of such relativistic velocities in our direction? Might they reach us from even more distant active galaxies? The origin of the charged particles in cosmic-rays from such events is generally accepted, but the possibility of more substantial pieces of high-energy matter is more speculative. See for example:

https://www.sciencemag.org/news/2014/07/physicists-spot-potential-source-oh-my-god-particles

Astronomers identify two classes of stars that explode as supernovae. The first is 'core collapse' (types II, 1b and 1c); these are massive stars that leave a compact remnant of some kind, a pulsar or a black hole. Typical examples of this type would be the *Crab Nebula* in the constellation Taurus and its central pulsar; or *SN1987A* which exploded in the Large Magellanic Cloud in 1987; the most recent that has been observed in our Galactic neighbourhood. The Crab pulsar is observed to rotate 33 times per second, and the surrounding nebula is expanding at about 1500 km/s. Another example would be *Cassiopeia A*, a strong X-ray source, which should have been first-visible from Earth around 340 years ago but went unrecorded. The remnant is seen to be expanding away from a central pulsar at: *"up to 31 million miles per hour* [13800 km/s] *(fast enough to travel from Earth to the Moon in 30 seconds!)"*. For more information about Cassiopeia A, including a Hubble photo, see:

https://svs.gsfc.nasa.gov/30951

The second class of supernova is a 'thermonuclear explosion' (type 1a); less powerful but likely to be more common. We may compare these to a vast nuclear bomb. They are believed to be white dwarf stars that have accreted a shell of matter, usually pulled from a close companion star. When they reach maximum size (Chandrasekhar limit) they must contract further to a neutron star but this can cause a thermonuclear runaway that completely disrupts the progenitor star. They leave no central remnant behind and by definition they should all be of a similar energy. Tycho's supernova SN1572 is one example, whose visible remnant has been variously calculated to be expanding asymmetrically at between 5,000 and 9,000 km/sec. [1]

https://www.nasa.gov/mission_pages/WISE/multimedia/gallery/pia13119.html

Kepler's supernova of 1604 may be another instance of this type of explosion within a close-binary system but its cause remains uncertain – a possible merger of two white dwarfs; see:

https://phys.org/news/2018-08-kepler-supernova-explosion-survivors-left.html

SN1604 is described by astronomers as having been 'unusually powerful'.[2] However, the estimated rate of expansion is comparatively low at 4200 km/s [3]. The recently discovered SNIa G1.9 + 03 may be another example of this type.

Astronomers observe that stars don't just explode without warning. Massive stars approaching their crisis will be surrounded by a dense planetary nebula ejected by novas in the later phases of their evolution. Within and beyond this will be the cloud of planetesimals, asteroids, meteors and comets that have circled the star ever-since it formed, equivalent to the Oort cloud around our Sun. These bodies are too small be detected but we know they must be present. Consider how many billions of icy comets and small asteroids orbit our own sun, completely unseen until one of them streaks through the sky as a comet.

A prime example of an 'overdue' supernova is the star *Eta Carinae*, which survived a nova explosion observed in 1843 (a supernova 'imposter'). The ejected matter is seen to be travelling at 32 million kilometres per hour! Astronomers cannot explain how the central star survived the nova as it is burning heavy elements in its core, its hydrogen fuel – long gone. It makes a startling image in the Hubble photographs.

https://www.universetoday.com/142734/hubble-has-a-brand-new-picture-of-the-massive-star-eta-carinae-it-could-detonate-as-a-supernova-any-day-now/
https://astronomynow.com/2018/08/03/astronomers-stunned-again-by-eta-carinae-the-star-that-will-not-die/
https://www.youtube.com/watch?v=cDOHshyLHdY

Fascinating as these spectacular examples are, it is probably the closer and less spectacular supernova remnants; those we can no longer see, which offer the more likely danger to Earth.

The final collapse of a massive star occurs within a fraction of a second as it exhausts its silicon and lighter elements and has only iron remaining in its core. This cannot release further energy from nucleosynthesis to support itself and can only collapse down to incredible density as the iron is crushed to a rapidly-spinning ball of neutrons – a neutron-star.

You may wonder how a collapse can cause an explosion. Various mechanisms are proposed and need not be discussed in detail here as we are more concerned with the remnant rather than the explosion. Most often discussed is that the collapsing matter 'rebounds' from the dense nucleus and collides with the in-falling matter triggering nuclear synthesis; and that the resulting burst of radiation and neutrinos would blow away the shell of the star. However another theory considers that the principle mechanism could be the rapid spinning-up of the core as it forms the central pulsar or black-hole; this throws-out the neutron-rich matter and the collapsing shell, in what is sometimes termed a 'sausage instability' (mass-shedding via outgoing spiral arms of matter). [4] Remember this all occurs within a fraction of a second! *A naive illustration would be to swing a string of sausages round your head – it is the fast-moving outer links that are most likely to fly-off at a tangent!* These ejecta collide with the in-falling stellar envelope and triggers further synthesis of heavy elements and vast release of radiation that we observe. The energy is carried away in a shock-wave that sweeps up everything surrounding the star and creates the visible nebula. The expansion of the ejected remnant then proceeds as the supernova fades.

We may propose that when the shocked ejecta reaches the outer shell of meteors, asteroids and icy-comets surrounding the exploding star then these would be disrupted into smaller fragments. Usually astronomers consider total disruption to gas and dust; and that the expansion ceases when the mass of the surrounding shell equals that of the unshocked interstellar medium. This may be so for most of the mass. However, bodies of optimum size and distance from the explosion may not be disrupted completely. If the explosion has enough energy to disrupt an asteroid then it must also be sufficient to propel away the fragments. Much depends upon how fast the collapsar is spinning. There is nothing in the interstellar vacuum of space that could then impede this shell of ejected meteors from expanding to infinity unless they encounter the gravity of another massive body (i.e. a star or planet).

[A crude analogy for the non-astrophysicist: consider your garden blower as it easily blows-away the sand and leaves; the blower is strong and the leaves are light. However, if someone throws a pebble at you could you deflect it away with your leaf-blower? I don't think so! How strong would your blower need to be; and how small must be the projectile for the blower to have any effect? How hard would you need to blow to completely disrupt the pebble?]

A further source of fast ejecta from a supernova may originate from neutron-matter expelled from the dense core as it spins-up. Astronomers theorise that such matter should very rapidly decompress to heavy nuclei (*r-process* elements) and iron. Again, these are likely to clump rather than persist as dust and gas. Sometimes trails can be observed within supernova remnants, which are interpreted as fragments from the core travelling through the gaseous remnant. For each of these there must be many smaller unobservable clumps. What is to impede these ejected pseudo-comets from escaping to great distances?

A complex collection of supernova remnants of various ages are found in the Vela nebula, revealed in a striking composite picture released by NASA.

https://apod.nasa.gov/apod/ap190110.html

You may also like to read:

http://blair.pha.jhu.edu/hstvela/hstvela.html

If we take as an example the paper by Loeb-et-al; the authors consider the V-shaped wakes observed within the Vela supernova remnant that are loosely estimated to be moving away at around 3000 km/s. [5] They demonstrate that these could not be the ejected former planets of the parent star, as these should be totally disrupted and would long-since have faded from view. At the point where the fragments cease to be observable, the professional astronomers lose interest; the planetary fragments are assumed to vanish into the supernova remnant. The disrupted planetary fragments, travelling at even a fraction of the remnant's expansion velocity, are still very fast compared to solar comets.

Rather, the authors consider that the observed high-velocity fragments are more likely to have been formed in the dense core and then ejected as the central remnant spins-up (an asymmetric gravitational collapse). The neutronic matter re-forms into heavy nuclei, principally iron and nickel – but unlike a solar iron meteorite these should also contain short-lived isotopes of heavy elements and iron-60. The authors consider that there must be "many more, smaller fragments" that we cannot detect. [6] Again the professional astronomer seems to lose interest if the bodies are too small to be detectable with telescopes. In fact, the diffuse shell must continue to expand long after the remnant has faded and they will one-day reach us. How many other expanding shells of older supernovas are out there, whose nebulae have faded from view? Their 'comets' are still on their way towards us, or have already passed through the solar system.

Until quite recently, astrophysicists believed that all the elements heavier than iron were created in supernova explosions. However, it is now accepted that even the high density of a supernova core-collapse is not dense enough for the *r-process* to create these elements. To produce these requires the merger of two neutron-stars, or perhaps a neutron star and a white dwarf. These events generate intense magnetic fields that fling jets of matter away from both poles of the collapsing star at near-relativistic speeds. These are also now believed to be the source of the *gamma-ray bursts* observed in distant galaxies; the bursts that are detected come from chance events where the polar jets are pointed directly at us. It is reasonable to expect that iron-rich meteor fragments might also be accelerated away by these intense magnetic fields.

The next problem to consider is how often we may expect to encounter the diffuse shells of fast-moving meteors and comets; and how may we observe them in order to prove their existence? In our historical records, visible supernovae have occurred about ever 300-500 years on average, the most recent being that observed by Kepler in 1604. In all, there should be about two events per century for the Milky Way galaxy, which would give the above average for our observable region. [7] Events on the far side of the galaxy should represent no danger to us. Therefore, the expanding shells of ancient ejecta should arrive at the solar system with a similar regularity to the observed supernova events. Between these episodes we are unlikely to encounter supernova ejecta and it may be that no such encounter has occurred since the advent of telescopic astronomy. In most events, they may be expected to rapidly pass through the solar system without hitting a planet. Therefore we may postulate a collision-event perhaps every few thousand years; really energetic axis-tilting events should be even less common by a multiple of 180. Without precise observational data to cite, the speculation flag will inevitably be raised!

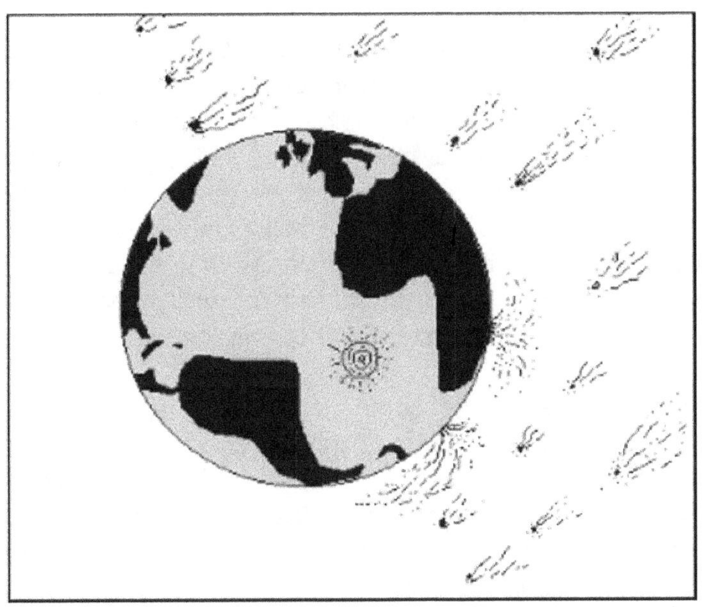

A flux of small-fast-comets arrives at the Earth.
This illustration was figure 10.1 of *Under Ancient Skies*.

When astronomers and popular commentators discuss the dangers from supernovas they tend to seek those that may explode within say, 500 light-years – the solar neighbourhood; such as the bright star *Betelgeuse* in Orion. The popular view is that as long as the supernova is more than say, 50 light-years away then we should be safe. The focus falls upon the gamma-ray burst or the intense flash of cosmic radiation and neutrinos that might reach us at the same time as we see its light; some others may consider the 'bubble' of charged particles, travelling typically at a fraction of light-speed that should reach us a few hundred years later. By contrast, many thousands of years may pass for the hyper-velocity solid ejecta to reach us from stellar explosions.

The greatest danger to Earth therefore comes from a 'shock-front' of these small fast 'comets' from ancient supernovae travelling at velocities 5-10,000 km/sec or perhaps even higher velocities. Compare this to the maximum for solar comet impacts of only 72 km/sec. A little calculation will show that at the above velocities, a shock-front of small fast debris like this would cross the orbit of the Earth in less than a day and pass-through the entire solar system within a week; probably too fast to form an observable tail. It must be remembered that we consider here objects perhaps as small as meteorites travelling at the hyper-velocities discussed. Unless they hit a planet we would probably never see them. We can only hope that they pass through the solar system without hitting us, undetectable and undetected.

However, we may get some advance warning. The charged particles from the supernova, travelling at a fraction of the speed of light, should reach us well in advance of the solid bodies; we should therefore observe a change in the background cosmic ray flux. Perhaps astronomers could then calculate their point of origin and predict when the dangerous 'comets' will arrive. Then what should we do?

We should also consider what the impact of a small-fast meteor on the Earth might look like; take as an example, a football-sized icy-comet, perhaps with a solid core, travelling at the ejected supernova velocities discussed above. The composition is irrelevant, ice would be just as destructive as iron; it is the velocity that supplies the kinetic energy not the mass. There is no ideal research to cite here as such impacts have not been much considered by specialists. The best one can suggest is that it would drill deep into the crust, more like a bullet-hole than an impact crater. The impact would throw-out behind it a tail of terrestrial rock far exceeding the small mass of the projectile, which would then dissolve deep within the earth. The best research we can consider would be the various theories for the formation of *tektites*, where we find strewn fields of these tiny impact-ejecta that are considered to be shocked glass formed from melted sedimentary rock as it re-enters the atmosphere. [8] Other than the tektites very little evidence would remain after a few thousand years; no crater, no tell-tale layer of Iridium, no evidence of a dust veil. We might however expect to find traces of heavy elements, and rare radioactive isotopes around the impact site, – if we could find one!

The consensus seems to be that very special and seldom-encountered high-energy events are required to explain the composition of tektites. For further information, the reader may like to pursue the following links: a basic introduction plus a more scholarly article:

http://earthsci.org/space/space/tektites/tektites.html
http://www.jsg.utexas.edu/npl/outreach/tektites/

A high energy impact event therefore has the capacity to supply an *impulse* of kinetic energy in the direction of its travel. Depending upon the size, direction and velocity, this has the capacity to change the Earth's angular momentum; and to excite the *Chandler Wobble*. Without 'hard' evidence then all we should expect to see in the geological record of the recent-past is evidence of the transient wobble and a pole shift; perhaps: a change of obliquity and a glitch in the length of the day. We might expect these effects *to be of the order of arc-seconds or arc-minutes at most*, but enough to be noticeable. This would manifest as sea-level and climate oscillations in the recent geological record; and in transitions from one stable regime to another.

However, *it must be stressed again* that in order to change the Earth's angular momentum by any significant amount, say, measured in full degrees or arc-minutes of latitude, then the impactor must arrive at a very high velocity perhaps approaching half the speed of light; and it would also have to be very small were it not also to cause a mass extinction event. To create impactors of such high energy we may therefore narrow down the source to those smaller chunks of heavy elements and iron ejected from the core of a supernova or star-merger event during its collapse. The only other source powerful enough might be the relativistic jets ejected from the poles of the collapsar during its rapid contraction phase; and one would have to be pointing in our direction from an ancient event. If it were closer then we could not escape the even more destructive effects of the gamma-rays! If something like this did reach us during Earth's recent prehistory then astronomers should one-day be able to identify a candidate progenitor star, which should by now be detectable as a black-hole.

Unfortunately, evidence of sea level and climate oscillations in recent prehistory can also be explained by other causes; and so without a clear worldwide pattern the sceptical geologists and climate specialists will not be convinced of catastrophic events from space. Astrophysicists are more open-minded; they are accustomed to discussing hypothetical phenomena that they can't touch; but geologists: they study rocks and they drill holes; it can be difficult to persuade them to look upwards.

References:

1)   Pilar Ruiz-Lapuente et al (2018) Tycho's supernova: the view from https://arxiv.org/abs/1807.03593

2) Pilar Ruiz-Lapuente et al (2018) No Surviving Companion in Kepler's Supernova, *The Astrophysical Journal* DOI: 10.3847/1538-4357/aac9c4 https://phys.org/news/2018-08-kepler-supernova-explosion-survivors-left.html

3) Vink, J. (2008) *The Kinematics of Kepler's Supernova Remnant as revealed by Chandra* https://arxiv.org/abs/0803.4011

4) A. Loeb, F.A. Rasio, J. Shaham (1993) *Ejection of Fragments in Supernova Explosions*, page 3; https://arxiv.org/abs/astro-ph/9405071

5) ibid, page 1

6) ibid, page 6

7) Colgate, SA ((1971*The velocity and composition of supernova ejecta* , page 74 https://ntrs.nasa.gov/archive/nasa/casi.ntrs.nasa.gov/19720004110.pdf

8) Koeberl , C. (1994) *Tektite origin by hypervelocity asteroidal or cometary impact: Target rocks, source craters, and mechanisms.* in B.O. Dressler, R.A.F.Grieve, and V.L. Sharpton, eds., pp. 133–152, Large meteorite impacts and planetary evolution. Special Paper no. 293. Geological Society of America, Boulder,Colorado. https://www.univie.ac.at/geochemistry/koeberl/publikation_list/095-Tektite-origin-by-impact-GSA-SP293-1994.pdf

This article was originally published in 2018 as an interactive web-page at: https://www.third-millennium.co.uk/dangers-from-ancient-supernovas

2   The Neolithic Calendar

Summary: *Plato's Atlantis myth holds a passing reference to a religious festival held by the ancient kings at alternate 5 and 6 year intervals. This may be evidence of an ancient calendar unrecognised even by Plato. The author has reconstructed this calendar and compares it with extant evidence of the calendar of the Druids. The preservation of a practical calendar within the Atlantis myth suggests that it has come through from a genuine ancient source; and that other aspects of the story, including the references to a flood catastrophe in the Atlantic, may be historical.*

For students of catastrophism in human prehistory, Plato's story of Atlantis will always be a primary source text. According to Plato, the myth of an island in the Atlantic Ocean, which disappeared beneath the waves in a single day and night, was an ancient Egyptian story transmitted to Greece, via Solon, who visited Egypt at some time around 590BC.

Classical scholars however, insist on treating the story as the invention of Plato himself and would seek to compare it only with other areas of Greek mythology. The possibility that it might truly have come from Egypt; and that it may contain a memory of a real flood catastrophe that befell the Atlantic coast of Europe, is left to more unconventional investigators to pursue. If it were possible to prove that one part of the story is true, then it would strengthen the arguments that other parts of it may be historical; and that it may hold a memory of a real ancient cataclysm.

When researching mythology, the present author's method is to look for 'mythological fossils'. These are pieces of detail that are not strictly needed to tell the main story. They may be references that can be independently checked, or which may have survived unchanged through generations of oral repetition of the myth.

One such vital piece of evidence is hidden within Plato's *Critias*.[1] Here he tells us that, at a remote period, before the catastrophic sinking of the island, the kings of the various parts of their empire gathered together at the temple of Poseidon:

...every fifth and every sixth year alternately, thus giving equal honour to the odd and to the even number.[2]

I first investigated this statement in my book *The Atlantis Researches* in 1995.[3] Why did Plato need to introduce these alternated five and six-year periods into his narrative? A fictional storyteller could simply have said that the ancient kings gathered together every few years to discuss policy. Why bother to be so precise?

It is common sense that in order for kings from the outlying provinces to gather on a prescribed feast day they would all need to be using a similar calendar and be able to measure the year to sufficient accuracy. We have here a piece of numerate evidence that can be checked by science. We can investigate whether it is possible to construct a real calendar that works on the principle of alternated five-year and six-year periods.

There is no reason to believe that such a calendar is either Greek or Egyptian, The Greeks of Solon's day, used a lunar calendar based on an eight-year cycle, the *Octaeteris*. Egyptian calendars, as understood by Egyptologists, offer no evidence of either five or six year devices. A cycle of five-plus-six years implies a lunisolar calendar based on an *eleven-year* intercalation cycle, requiring four intercalary months. It is important to appreciate that 11-years is a *real* intercalation cycle; indeed it is slightly better for the purpose than the Octaeteris.

The importance of intercalation cycles is that a precise number of months and years will repeat, with the smallest cumulative error, to reconcile the lunar year of 354 days (12 x 29.5-day lunar months) with the solar year. The insertion of intercalary months requires that some calendar years must have 13 months. The half-day in the lunar cycle requires that all calendars must combine both 29-day and 30-day months; and since even this is not exact, a variable month is always needed so that sun and moon can be regulated according to easily-understood rules.

The accuracy of various intercalation cycles

intercalation cycle (y)	intercalated months	synodic month	no of months (x)	days in x months	tropical year	days in y years	discrepancy (days)
3	1	29.530589	37	1092.6318	365.2422	1095.7266	3.094807
8	3	29.530589	99	2923.5283	365.2422	2921.9376	-1.590711
11	4	29.530589	136	4016.1601	365.2422	4017.6642	1.504096
19	7	29.530589	235	6939.6884	365.2422	6939.6018	-0.086615
22	8	29.530589	272	8032.3202	365.2422	8035.3284	3.008192
30	11	29.530589	371	10955.849	365.2422	10957.266	1.417481
57	21	29.530589	705	20819.065	365.2422	20818.805	-0.259845
76	28	29.530589	940	27758.754	365.2422	27758.407	-0.34646
84	31	29.530589	1039	30682.282	365.2422	30680.345	-1.937171

No calendar is known to have been based upon the 11-year intercalation cycle. The 19-year cycle was introduced to Greece from Babylon by Meton in 432 and forms the basis of modern Jewish and Muslim calendars. A 19-year cycle is out by only about two hours, whereas an 11-year cycle would be out by more than one and a half days.

So, if the Atlantis myth were merely Plato's fiction, then why would he go to so much trouble to invent a completely new calendar based upon a real eleven-year intercalation cycle? Moreover, why does Plato not mention the five-plus-six-year calendar in any of his other astronomical works? I believe that Plato never recognised this detail for what it is. It is evidence of a *real* ancient European calendar, carried through as a 'mythological fossil', surviving intact through language changes to Egyptian, Greek, and on into modern English. So far as I know, no researcher has investigated this until I did so in *The Atlantis Researches*.[4]

The same mythological fossil may be found within one of the songs of the Finnish epic *Kalevala*.[5] Here we find a story about the maid Marjatta who laments the passage of a long period of years as, "for five or six of summers". The use of this five and six year formula in Finno-Ugrian oral tradition gives us a strong clue that the 11-year cycle has its origin in prehistoric Europe.[6]

Another place where we encounter a five-year calendar cycle is in Celtic Gaul. In Julius Caesar's commentaries of his Gallic wars, between the years 58 and 51 BC, we find a brief account of the activities of the Gaulish Druids. Caesar apart, we find only a few passing references to their science in the works of other classical writers such as Pliny. Caesar tells us that the Druids were capable astronomers who debated the size of the Earth and the movements of the planets.[7]

The Romans ruthlessly suppressed the Druids in Gaul, especially during the reigns of Tiberius and Claudius and we must presume that they also persecuted them within their British province. Indeed the Druids' continuing influence in Gaul may have been a principal motive behind the Claudian conquest of AD 43. Caesar again, tells us that the order originated in Britain; and that a Druids College existed there.[8]

Evidence of how the Gaulish calendar operated is preserved in the Coligny calendar, which was discovered in a vineyard near Bourg-en-Bresse in 1897. The extant calendar consists of several fragments, constituting some three-fifths of a bronze tablet, which appears to have been deliberately broken-up and buried. The reconstituted calendar shows annotation in the extinct Gaulish language, using the Roman alphabet; and this has been used to date it approximately to the reign of Augustus. The mechanism of the calendar however, shows no Roman influence and must therefore belong to the final period of native Gallic culture. A less well preserved fragment found near Villards d'Héria is thought to date from the second century AD.[9]

Although most commentators on the Coligny Calendar have treated it as a purely Gaulish artefact no older than the Roman era, it is important to appreciate that all calendars evolve gradually from ancient roots. The first-century Gaulish Druids did not invent the Coligny Calendar. Take as an example the Gregorian calendar that hangs on your own wall; the days and weeks are pagan Germanic; the months are named from Roman months that predate Caesar's reforms and must be at least 2500 years old! The Coligny fragment may similarly hold evidence of astronomy that was performed in Western Europe thousands of years before Roman times. Caesar's comment that Druidism was 'found existing in Britain' would suggest that we should look to the British Isles for the origin of the calendar.

The Coligny tablet holds notation for five years; set out in sixteen columns. Each column holds the notation for four named months; the exception being the first and the ninth columns, which hold an intercalary month, followed by two normal months. The intercalations are therefore positioned at the beginning of the five-year cycle and in the middle of the third year. Hence, it may be seen that the intercalations were spaced at intervals of two-and-a-half years.

With a single exception, the length of each month is always either 29 or 30 days. Each month is marked-out in days, listing their various festivals; and with a peg-hole to mark each date. Since the average period of the lunar month is 29.53 days, any lunar calendar must incorporate alternating 29 and 30-day months. Furthermore, one of the months must also be variable to regulate the correspondence with the real Moon. The 30-day months are each suffixed as 'good', whereas the 29-day months are styled 'not-good'. As the month Equos ('horse-month') was similarly 'not-good', this indicates that it was not considered an ordinary thirty-day month and scholars have concluded that it was the variable month. It may have alternated in length between 28 and 30 days.

The extant fragment contains notation for only sixty-two months and covers a period of approximately five solar years. Sixty-two lunar months contain 1831 days; whereas five solar years require only 1826 days. The sixty-two months exhibit notation for a possible 1835 days. Therefore, if the Coligny fragment were to constitute the entire calendar then, by strict lunar reckoning, it would suffer a cumulative error of four days in every five years; and an error of nine days by solar reckoning. A five-year intercalation cycle simply does not work! Such a huge error defeats the purpose of the intercalary months. There must therefore have been a correction mechanism or another, missing, cycle alternated with this one, in order to pull back the divergence.

For a calendar to be of use to the general population, it must be organised according to simple rules that everyone will remember. The best way to do this would be to use one of the self-correcting intercalation cycles. A theory that the Celtic calendar was based upon the nineteen-year Meton cycle was proposed by Fotheringham and Rhys, some of the first scholars to investigate the Coligny calendar.[10] However, there is no historical evidence that such a cycle was ever used by the ancient Celts, either in Gaul or in the British Isles. Later commentators such as Olmsted have proposed a theoretical 25-year cycle, but he suggests that this replaced an earlier 30-year cycle. However, while there is evidence for a five-year festival, there is no textual evidence for the existence of a 25-year cycle in ancient Gaul.[11]

The reconstruction given here is based upon the principle of alternated five and six year periods and preserves all the known rules of the Coligny fragment. Within the six-year cycle the intercalary months are spaced every 3 years, to give the required 4 intercalations in 11 years. An extra day has been added to the variable month in the first eleven-year cycle and two days to the second eleven-year cycle. One consequence of this configuration is that the start of a month falls behind the real Moon by 3 days every 22 years. After 44 years it would therefore give the situation described by Pliny, who tells us that the Druids' month began on the sixth day after new moon.[12]

If this 3-day slippage is designed into the calendar then the lunar cycle is held naturally in balance with the solar cycle over twenty-two years, with a discrepancy of only about thirteen minutes. This compares most favourably with the two-hour discrepancy of the Meton cycle over nineteen years. The arrangement shown in the reconstruction is at least an improvement on the Greek *octaeteris*.[13] Certainly, it is greatly superior to the unreformed Roman calendar. A further advantage is that a simple cycle of 5 + 6 years repeats indefinitely, with no complex rules to remember.

However, if the eleven-year cycle is required to repeat with the 30-year Druid ages then a long-cycle of 330 years (30 x 11) must elapse before the two cycles will mesh. In my earlier work I suggested that a double Druid cycle of 60-years may have been employed – and indeed many such reconstructions are possible without violating the rules of the extant 5-year cycle. However, I am now satisfied that the simple 5+6 year formula given by Plato, repeated indefinitely, is all that is required.[14]

The present author's hypothesis would suggest that, in addition to the extant five-year cycle of the Coligny calendar, there should also be a 'lost' cycle of six years. However, until hard evidence of a missing six-year cycle comes to light from somewhere in the Celtic regions, there can be no conclusive proof that the Coligny Calendar actually operated this way.

The mechanism proposed is therefore an eleven-year lunisolar cycle of 4017/4018 days that was allowed to wander alongside a thirty-year cycle based on the observation of Saturn. This cycle actually requires less intervention than the Metonic cycle over two eleven-year cycles. The lunisolar difference over eleven years is almost exactly a day and a half, so doubling this:

Sun:	2 x 4017.6642	=	8035.3284 days
Moon:	2 x 4016.1601 =		8032.3202 days
Difference =			3.0082 days

This slippage might explain why the Celtic months began on the *sixth* day of the moon at the era when Pliny recorded it.

Thus we see revealed the remarkable accuracy hidden within Plato's Atlantis calendar. The calendar as reconstructed here is beautifully simple. Given the realities of astronomy, constrained by the extant Coligny fragment, we can be confident that any real calendar based upon alternating five- and six-year periods must work in a similar way. I would invite any scholar to put this mechanism into their own spreadsheets and satisfy themselves that this so.

The mention of this 5+6 year formula within Plato's narrative *is too great a coincidence* for it to be mere fiction. This passage has always convinced the present author of the authenticity of the Atlantis myth. The preservation of the 5-year cycle by the Druids, and its mention independently in a Finno-Ugrian myth, suggests that a real calendar of this kind may have been used by the kings of ancient Europe just as is recorded in the *Critias*. This further raises the question of whether other aspects of Plato's Atlantis narrative deserve more respect as a record of history.

Next Page:

A reconstruction of the 5+6 year cycle over 22 years as originally published in *Under Ancient Skies*. Other arrangements for the days per month are possible as long as the result conforms to the lunisolar cycle and the known arrangement of the Coligny calendar.

Year No.	Inter-calary month	Month 1 SAMON (JUN)	Month 2 DVMANN (JUL)	Month 3 RIVROS (AUG)	Month 4 ANAG'NT (SEP)	Month 5 OGRON (OCT)	Month 6 CVTIOS (NOV)	Inter-calary month	Month 7 GIAMON (DEC)	Month 8 SIMIVIS (JAN)	**Month 9 EQVOS (FEB)**	Month 10 ELEMBIV (MAR)	Month 11 EDRIN (APR)	Month 12 CANTLOS (MAY)	Days/ year	Days/ period	Mths/ period
1	30	30	29	30	29	30	30		29	30	**30**	29	30	29	385		
2		30	29	30	29	30	30		29	30	**29**	29	30	29	354		
3		30	29	30	29	30	30	30	29	30	**30**	29	30	29	385		
4		30	29	30	29	30	30		29	30	**29**	29	30	29	354		
5		30	29	30	29	30	30		29	30	**30**	29	30	29	355	1833	62.071
6	30	30	29	30	29	30	30		29	30	**29**	29	30	29	384		
7		30	29	30	29	30	30		29	30	**29**	29	30	29	354		
8		30	29	30	29	30	30		29	30	**29**	29	30	29	354		
9	30	30	29	30	29	30	30		29	30	**29**	29	30	29	384		
10		30	29	30	29	30	30		29	30	**29**	29	30	29	354		
11		30	29	30	29	30	30		29	30	**29**	29	30	29	354	2184	73.957
12	30	30	29	30	29	30	30		29	30	**30**	29	30	29	385		
13		30	29	30	29	30	30		29	30	**29**	29	30	29	354		
14		30	29	30	29	30	30	30	29	30	**30**	29	30	29	385		
15		30	29	30	29	30	30		29	30	**30**	29	30	29	355		
16		30	29	30	29	30	30		29	30	**30**	29	30	29	355	1834	62.105
17	30	30	29	30	29	30	30		29	30	**29**	29	30	29	384		
18		30	29	30	29	30	30		29	30	**29**	29	30	29	354		
19		30	29	30	29	30	30		29	30	**29**	29	30	29	354		
20	30	30	29	30	29	30	30		29	30	**29**	29	30	29	384		
21		30	29	30	29	30	30		29	30	**29**	29	30	29	354		
22		30	29	30	29	30	30		29	30	**29**	29	30	29	354	2184	73.957

First 11 year cycle: 1833+2184 = **4017** days
Second 11 year cycle 1834+2184 = **4018** days

Totals of 22 year cycle = **8035** **272.091**

Notes and References

[1] Plato, Critias, 119 (translation by Benjamin Jowett)

[2] Within the *Timaeus*, Plato says that the empire extended to the whole island and also to the continent: Europe as far as Tyrrhenia and Libya. However, he refers to only one specific sub-king who ruled over a peninsula of the island. By analogy with the later empires of Rome and Napoleon's Europe, we may posit that tributary kings also ruled over the conquered territories of Europe.

[3] Dunbavin, P. *The Atlantis Researches – the Earth's Rotation in Mythology and Prehistory*, Third Millennium Publishing, Nottingham (1995)

[4] Dunbavin , P. *Atlantis of the West – The Case for Britain's Drowned Megalithic Civilisation*. Constable & Robinson, London (2003)

[5] W.F. Kirby (London 1907) *Kalevala: the Land of Heroes*, (republished by the Athlone Press, 1985); see Runo L, p 634.

[6] It should be recorded that when the author attempted to publish a similar article on the 5+6 year calendar in a journal of astronomical history, the *Kalevala* reference was dismissed as 'a patently solar reference' having no place in a discussion of a lunisolar calendar. Indeed, the very mention of Plato's *Critias* as the primary source for the calendar evidence produced a dismissive negative reaction. Hence the article was revised for publication elsewhere.

[7] Caesar, The Gallic War, VI.xiii-xiv; VI.xviii.

[8] Caesar, The Gallic War, VI,Xiii

[9] Duval, P.-M & Pinault, G Recueil des *Inscriptions Gauloises, 3: Les calendriers (Coligny, Villards d'Heria)*. Paris: Supplement a Gallia 45 (1986)

[10] Mac Neill, E. On the Notation and Calligraphy of the Calendar of Coligny, *Eriu*, X, pp1-67 (1928)

[11] Olmsted, G. *The Gaulish Calendar*, Dr Rudolph Habelt GMBH, Bonn (1992)

[12] Pliny, Natural History, XXX.xiii

[13] Geminus, Elementa Astronomiae, C.8.

[14] Dunbavin, P. *Under Ancient Skies: Ancient Astronomy and Terrestrial Catastrophism*, Third Millennium Publishing, Nottingham (2005)

3　On the Coligny Calendar and the Neolithic Calendar in Plato's *Critias*

An article published in SIS Review similar to the previously unpublished 2006 paper.

Summary

Evidence of a calendar within the Critias of Plato; and extant evidence of the calendar used by the Druids of Gaul can both be reconstructed according to an eleven-year lunisolar cycle of 5+6 years and may in fact be the same calendar. This short article summarises the author's previously published work on these subjects.

Introduction

The existence of astronomical alignments at Neolithic monuments such as the stone circles, once itself controversial, is now virtually accepted; somewhat intuitively these may be associated with various Celtic seasonal festivals and yet without any real understanding of how their calendar worked. Did they have a system of calendar dates and an era such as we do?

Recent DNA studies have largely overturned the old notion of a wave of Iron-Age Celts invading Neolithic Britain and Ireland. Neither may we any longer talk vaguely about a 'Celtic' calendar as if continental invaders had introduced their religion and culture, including their calendar, at some convenient period just prior to the Roman invasions. When we read Julius Caesar, who says that the Druids and their culture were 'found existing' in Britain, then we need to take such a statement by a contemporary chronicler far more seriously. We may now only use the term 'Celtic' as a loose linguistic grouping.

It is not my intention here to go too deeply into source references in what is intended as an interesting article for a general audience rather than a scientific paper. I explored these themes in much greater detail some years ago in my books *Atlantis of the West* and *Under Ancient Skies* and in a further unpublished paper; where all the source references may be found for anyone who wishes to verify my conclusions.[1] There is nothing here that I have not published previously.

The Calendar of Coligny

Clues to the calendar of the Gaulish Druids, and thereby of their astronomy, come from the "Calendar of Coligny" (Fig. 1), discovered in a French vineyard in 1897. It was part of a wall plaque clearly designed for display, similar to the annual calendar on your own wall. Its notation displays a five-year cycle of lunar dates. Conventional academic assessment has been inconclusive because a five-year period does not work as a lunisolar cycle. This has led to a conclusion that the Celts had only a poor knowledge of astronomy.

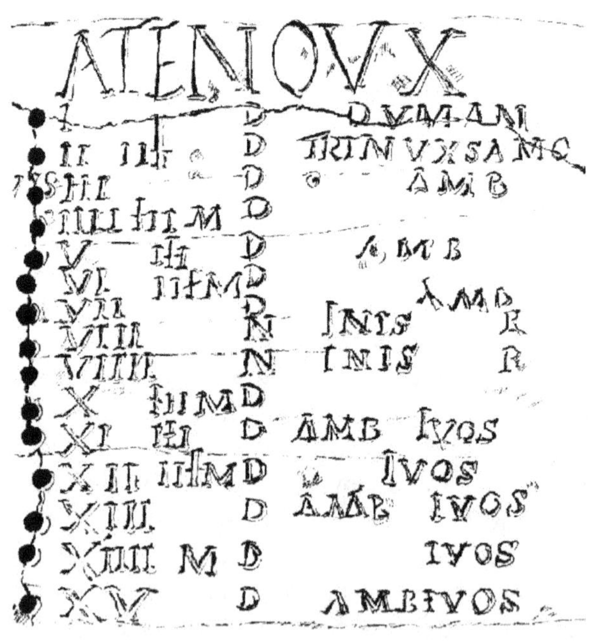

Fig.1. A fragment of the Calendar of Coligny

The structure of the calendar shows 62 months of 29 and 30 days arranged as follows:

```
30   30   29   30   29   30   30   .    29   30   29*  29   30   30
.    30   29   30   29   30   30   .    29   30   29*  29   30   30
.    30   29   30   29   30   30   30   29   30   30*  29   30   30
.    30   29   30   29   30   30   .    29   30   29*  29   30   30
.    30   29   30   29   30   30   .    29   30   29*  29   30   30
                                                      Total: 1831 days
```

The month called *Equos*, 'horse', had nominally 29 days and was variable according to rules that cannot be determined from the fragment alone. It may sometimes have been 28 or 30 days, so the 5-year cycle could have been anything between 1826 and 1835 days.

Other Ancient Calendars

Conservative academic opinion will only allow that the Calendar of Coligny was devised by the Iron-Age Druids in Gaul just prior to the Roman invasion. A moment's thought will show the problem here. The Gregorian calendar on your own wall was printed last year, but the printer did not devise it! It holds months named after Roman emperors and days named for Germanic gods that are thousands of years old; the structure is but a revision of the old Latin calendar. Therefore, we may assume that the devices on the Gaulish calendar similarly evolved from much earlier roots. This would take us back to the Neolithic and the era of the megalith builders.

The basis of any such calendar is a natural lunisolar cycle, which arranges the months so that they stay naturally in step with the seasons. Ideally, festivals should repeat at roughly the same season every year without requiring constant revision by priests and astrologers. In a solar calendar, such as our familiar Julio-Gregorian calendar, it is the solar year that is held and the lunar phases are allowed to appear as they fall. However, in a lunar calendar the months must begin strictly with a phase of the moon and be arranged as either 29 or 30 days long. Consider the following (*current astronomical data used here and subsequently*):

One lunar month =	29.530589 days
62 lunar months =	1830.8965 days
One solar year =	365.2422 days
5 solar years =	1826.211 days
Difference:	4.6855 days

As the month and year are not integral, the best arrangement is therefore to use a cycle where a multiple of lunar months is almost equal to a multiple of solar years. There is no precise equivalence, but some of the closest available are 8 years, 11 years and 19 years. The Greeks used the 8-year *Octaeteris*; the Babylonians used the 19-year Meton Cycle; but no culture is attested to have used an 11-year cycle.

In the Greek Octaeteris, 3 extra intercalary months were added to make the lunar years equate (roughly) to the solar years; in the Meton Cycle 7 extra months; and in an 11-year cycle 4 extra months would be needed; so all lunisolar calendars must employ some lunar years of 13-months in addition to the 12-month years. If these are evenly spaced then solstices and equinoxes never slip too far from their proper season. A good analogy here is Christian Easter, which shows us the opposite case of a lunar date wandering within a solar calendar, but always held in springtime.

However, the five-year period of the Calendar of Coligny is not a natural lunisolar cycle. 62 lunar months are 1831 days; 5 solar years are 1826 days. This is an accumulating 5-day slippage. And yet the notation shows provision for two intercalary months inserted every 2.5 years. *This suffices to tell us that the 5-year cycle is not the entire calendar.* The Intercalary months serve no purpose unless they were intended to hold the months in line with the seasons. So which cycle were they using? Was it 8 years? 11 years? 19 years? Or perhaps they used some longer cycle.

Another factor to consider is how to make the days add-up. Every calendar must employ a month of variable length inserted according to established rules. In the Julio-Gregorian calendar we add a day to February each fourth year and because this is still inexact the Gregorian reform added a century rule. In the Coligny Calendar we similarly see notation for a variable 29 day month. There is not enough evidence to say how the rules were applied – but we know that there *must* have been such rules or these extra days, like the intercalary months, would serve no purpose.

Other clues are offered by ancient writers who tell us a little about the Gauls; also about the ancient Britons; and of their priestly cast known as the Druids. Some other clues may come via myths and legends. If one may quote Pliny the Elder here: [2]

Mistletoe...is gathered with great reverence, above all on the sixth day of the moon (it is the moon that marks out for them the beginning of months and years and cycles of thirty years) because this day is already exercising great influence even though the moon is not half-way through its course
(Pliny, *Natural History*, XVI, 250).

The historian Plutarch also tells us that the Britons held their most important festival each thirty years as the planet Saturn returned to the constellation Taurus. [3] Consider:

One synodic period of Saturn = 378.09 days
29 oppositions of Saturn = 29 x 378.09 = 10964.61 days = 30.02 solar years

In my earlier reconstruction of the Coligny Calendar, I therefore also considered whether the Druids tried to base it on a precise thirty-year cycle held to Saturn's rhythm. So I investigated how the extant five-year 'Coligny' cycle could be alternated with a six-year cycle. This would allow the short-term calendar to mesh both with the thirty-year ages and also the 11-year lunisolar cycle.

This brings us on to the 5 + 6 year calendar intimated in Plato's *Critias*. To recap his Atlantis narrative, he tells us that the Egyptian priests remembered an island in the Atlantic that was struck by a geological catastrophe at a time just before the beginning of the Egyptian state. The ancient dynasty of kings who ruled this island and the Atlantic coastal regions would gather together *"every fifth and every sixth year alternately"* to discuss policy. If they were all to know when to meet then they must have used the same calendar; and Plato is giving us a clue how it worked.[4]

The Coligny calendar gifts us a detailed knowledge of how a real 5-year cycle was constructed. From this it is possible to work-out how a 6-year cycle *must* be arranged in order to make use of the 11-year lunisolar cycle.

In the Coligny fragment the intercalary months are evenly spaced at 2.5 year intervals; one at the start of the five year cycle; the other in the middle of the third year. If we reconstruct a six-year cycle using the same month layout, then the intercalary months should naturally be spaced at the beginning of the cycle and at the start of the fourth year. Therefore we may propose:

30	30	29	30	29	30	30	.	29	30	29*	29	30	29
.	30	29	30	29	30	30	.	29	30	29*	29	30	29
.	30	29	30	29	30	30	.	29	30	29*	29	30	29
30	30	29	30	29	30	30	.	29	30	29*	29	30	29
.	30	29	30	29	30	30	.	29	30	29*	29	30	29
.	30	29	30	29	30	30	.	29	30	29*	29	30	29

Total 2184 days

Here, for simplicity, I have assumed that the variable month always has 29 days, so accuracy is down to how you wish to organise the number of days that it has in each cycle; i.e. whether the calendar is to be held strictly lunar; strictly solar; of perhaps to follow Saturn.

These two cycles, alternated, give 4015 days, whereas 11 solar years require approximately 4018 days; so the variable month would need an extra 3 days to be added each 11 years. One cannot know for sure how the calendar rules actually operated, but it doesn't really matter *where* you put the extra days; so long as the number of days in the cycle add-up to your objective and this is fixed by astronomy.

So the 5+6 year cycle casually mentioned by Plato can produce a very accurate calendar and the reconstruction of the Calendar of Coligny yields a similar result. Is this a coincidence?

Decide for yourself. Of course the Neolithic calendar would not look exactly like that of Iron Age Gaul, it's a process of evolution, just as our Gregorian calendar has evolved from its original Roman roots.

It was this detail, hidden in plain sight within the *Critias*, that convinced me of the authenticity of Plato's narrative – but you must decide for yourself how much of his catastrophism you wish to accept alongside the description of ancient geopolitics.

Classical scholars, if they are to remain respectable, may only discuss Plato's Atlantis narrative as if it were a classical Greek story. They are not allowed to look beyond Crete, or the volcanic destruction of Thera, for the inspiration behind Plato's lost island; yet its own internal content states that it was based on history told to Solon, as recorded by the Egyptian priests. The story is *Egyptian*, not Greek; they recorded the organisation of an island in the Atlantic and of the Atlantic coastal regions. They tell us that this was contemporary with the earliest period of the Egyptian state (pre-dynastic). Egyptologists determine this was the late fourth millennium BC.

Archaeology tells us what was going-on along the Atlantic coasts of Europe in the late fourth millennium BC. This was the middle-Neolithic; a time when the people began to build megalithic monuments with calendrical alignments; in Britain, in Ireland, in Brittany and as far away as Malta. Is it therefore so far out to equate the two? Is it so unacceptable to say that if you have a calendar on your own wall with notations that are thousands of years old, then perhaps the first century Gauls also used a calendar that was thousands of years old?

My own research has taught me always to believe what the ancient sources actually say, rather than what academics think they are saying.

Notes and References

---

[1] *Atlantis of the West* (2003) and *Under Ancient Skies* (2005) Both books are now available again in Kindle editions and an unpublished paper intended for *C&C Review* is available (see previous), or via www.third-millennium.co.uk

[2] The Latin is genuinely ambiguous here; some translators give that the months actually began on the sixth day after new moon.

[3] This shows us that the Druids were observing the Synodic Period; the time required for a planet to return to the same position relative to the Sun as viewed by an observer on the Earth.

[4] If you doubt the importance of using a common calendar, then recall when, in 1805, the Russian Tsar sent his army to join with the Austrians before the Battle of Ulm, only to find that Napoleon had already defeated them! The Russians were still following the unreformed Julian calendar which by then was 11 days out of synchronization with the Gregorian calendar used by the Austrians.

Citation: Chronology & Catastrophism REVIEW 2018:2 pp 50-53

4   Throwing Planets Around

Summary

Between 1999 and revised up to 2018 a series of papers proposing the Planet-Z Hypothesis were published by a group of physicists and astronomers, including the eminent astronomer Willy Woelfli; these would seek to revise and augment the ideas of Immanuel Velikovsky with more realistic physics. This review seeks to constructively compare and contrast the strands of enquiry to date.

The Planet Z hypothesis is granted a greater degree of scientific credibility and exposure than it might otherwise attain, due to the eminence of the authors and its presence on the Cornell University online archive. Physics professors can perhaps cite Velikovsky without risking credibility – but no-one else could expect to have similar work published on such prestigious platforms.

The co-authors of the series of papers are *Professor emeritus Dr Willy Woelfli* and *Professor Dr Walter Baltensperger;* together with *Robert Nufer*, who describes himself as an amateur astronomer with a particular interest in comets. Here, to avoid repetition, I shall refer to them as 'the authors', or 'the co-authors' and cite the papers either collectively or by individual publication date. The background of the three authors is described in their own papers listed below (Robert Nufer's astronomy website is recommended reading). [1]

To attempt a summary of the theories, with apologies for brevity, would be as follows. The 1999 paper (later revised for v1 in 2018) introduces the concept of a periodic close approach by an additional planet as a possible explanation for Earth's climatic changes in the past few million years. This idea was further expanded in the 2002 paper by the same authors. The model simulates a planet in an eccentric Earth-crossing orbit as a principal cause of the Pleistocene ice ages between 3.5m years ago and its (apparent) end 11.5k years ago; this planet they term 'Z' to substitute for Velikovsky's rogue 'Venus'. The highly eccentric orbit of this hypothetical body allowed several close approaches during the Pleistocene Ice Age; during these encounters, especially during its final pass, the tidal forces deformed the Earth resulting in a pole shift that dragged the North Pole away from Greenland to its present location. The final close-encounter resulted in the break-up of Z which no longer exists. The age of ice is therefore considered to be over and we are back to the stable conditions that existed back in the Miocene epoch.

The hypothesis stresses that the Earth's obliquity was scarcely affected by the Pleistocene encounters and that the geographical pole shift was caused by a tidal bulge resultant from the close approach of Z, which altered the rotational balance. They then proceed to calculate the mass, structure and orbital properties of the transitory planet in order to cause the required tidal effects. In gradualist geological theories the trigger for ice ages is the drift of a continent over one or both of the poles; in a catastrophic event such a transition could take place more rapidly due to a pole-shift.

The co-authors comment on the doubtful physics underlying Velikovsky's original astronomy. Much stress is also placed upon his letters to Albert Einstein and the physicist's reply: *"Catastrophes yes, Venus – no!"* What a marketing coup that was for an author! But why did the co-authors find it necessary to commence from a base of 1950's science? [2]

The computer simulation reveals that the required tidal effects demand an extremely close approach by a planet with at least the mass of Mars (therefore truly a planet and not a comet). The repeated visits of the disintegrating Z to the inner solar system would have left a debris trail along its orbit, thereby triggering a cooling effect, due to the dust veil in space and in our atmosphere. Another feature of the model is that the close-approaches of Z and its dust veil are considered more influential upon ice age climate than the Milankovich effect. Since all trace of Z has now gone (save perhaps for periodic meteor showers) then it must have been either ejected or dissolved, or the fragments fallen into the sun. Such a break-up demands that in its final approach Z must have entered within the Earth's Roche limit – very close indeed!

The authors then offer a discussion of which facts become plausible or explainable by their model. The most obvious of these is the survival of mammoths in Siberia and the evidence that Siberia was never covered by an ice cap; Lake Baikal did not freeze and its unique fauna survives; the Himalayas were not ice-domed as were the Alps. These anomalies are not explainable by gradualist geological theories. The semi-tropical conditions at mid-latitudes during the 'interglacials' are also glossed-over in gradualist theories, as are the so-called 'Heinrich events' when the polar sea-ice apparently broke-up. The co-authors believe that their modelling can also explain such anomalies; it is only the conclusion of these climate fluctuations that has allowed human civilization to flourish.

The 2007 paper builds upon the earlier studies with more concise information and mathematics. A further summary of this additional detail would be as follows:

The eccentric orbit of Planet Z caused it to become extremely hot due to *'tidal work and solar radiation'*. The final close encounter produced a *'stretching deformation'* of the Earth that was the ultimate trigger for pole-shift. The hypothesis suggests that the survival of dwarf mammoths on Wrangel Island indicates a change of latitude and adds further comment that the Milankovitch effect *'cannot be understood'* without assuming a formerly lower-latitude for Siberia; also that it cannot explain why the same secular orbital factors did not apply back in the Miocene.

Planet Z must have been breaking-up before the final encounter, but the computer simulation, they admit, does not allow for this shedding of mass. However they conclude: *'...that Z had at least 1/10 of the mass of Earth...'* and remark: *'For example Z might have been a moon of Jupiter which got loose'* and *'Clearly, Z had to be hot and emitting material already before the polar shift'*. The authors then consider the lifetime of a cloud of particles from the disintegrating Z:

> *The orbits of the individual particles will be spread over rather vast space so that initially particles do not collide. Their lifetime is limited by Poynting-Robertson drag ... valid for circular motion* [see note 3 below].

They argue for traces of the dust cloud in ice-cores and that:

> *A first version of this paper was written under the impression, that a terrestrial origin of the inclusions in bore ice had been demonstrated.*

However this constraint only applies to the dust rather than any larger fragments residual from the final break-up. The final pass of the disintegrating Mars-sized Z was described in the 2002 paper:

> *We expect that the peak tidal force has to be about an order of magnitude larger* [than previous encounters]... *this brings the closest distance between the centres of Earth and Z into the range of 12,000 to 15,000 km. As a result, Z enters the Roche limit of Earth.*

Regarding the end of the Pleistocene they add:

> *We just note here that the last rapid increase of the temperature recorded in the polar ice data occurred at 11 500 ± 65 yr BP...Thus, it appears that the Younger Dryas, which begins at 12 700 ± 100 yr BP is younger than the polar shift event.*

A further paper in 2010 was devoted to the folk-memory of these events in various ancient sources. These references were added because the earlier papers gave rise to comment that Z should be remembered in the old traditions. Here they cite

Chinese myths of 'persistent daylight' such that people were unable to sleep; and the myth of twin suns in the sky. This glow is attributed to the 'hot' Z, which they equate with the Typhon comet as recalled by Pliny and other classical writers. A comparable legend from the Ojibwa of North America is similarly advanced.

Next they cite Herodotus who describes a memory of the sun rising in the west and setting in the east, as given to him by the Egyptian Priests of Hephaestus; together with their calculation based on a priest-list that the age of Egyptian civilization stretched back as much as 11,340 years. The authors cite Herodotus, as did Velikovsky before them, that during this period the sun reversed its positions of rising and setting twice without detriment to life in Egypt. The late Peter Warlow in *The Reversing Earth* would try to explain this by the Earth turning completely upside down on its axis due to a similar tidal influence. [3] The authors prefer to explain it as a memory of Z in the sky rather than of the sun.

The authors remark on the Biblical references in *Joshua* to the sun standing still in the sky; and the close pass of a fragment of the lost Z as described in the Book of *Ezekiel* (594 BC – hence long after the final break-up). Another Biblical citation comes from the Book of *Revelation*; and a related quotation from Josephus regarding a time of ancient cataclysm. Reference is also made to the events of *Exodus* as a memory of relevant events, but the seven-year famine of Joseph is not cited.

Next they discuss the Egyptian *Papyrus Ipuwer* (First Intermediate Period) as a memory of the terminal-Pleistocene pole shift. They would view the papyrus as a recollection of various catastrophes from much earlier times rather than a reference to contemporary events.

Plato's *Critias* and *Timaeus* are extensively quoted. Here the authors cite the date given by Plato (Solon) again based on a chronology supplied by Egyptian priests. Therefore, they follow the literal dating assumption that he describes events at the end of the Pleistocene c.12 000 years BP; and therefore that an organized state must have existed both in Egypt and Greece at that date

Further memories of the Flood, the authors continue, are to be found in the *Epic of Gilgamesh*; preferred here to a discussion of the Biblical Noah. This event they attribute to the tidal forces produced by the passing planet. The passage of Z is more directly recalled in the words of Ovid's *Metamorphoses*. This recalls the comet Phaethon, perhaps a fragment of Z, which caused burning air and deserts in the Egyptian Delta and elsewhere. The authors offer a table showing the changes of latitude required to explain these various recollections, based on the pole shift that they propose.

Of crucial importance is the apparent memory of a change to the number of days in the year from 360 days to our present experience. Here the authors draw again on Velikovsky and state:

> *Conservation of angular momentum grants that the direction of the Earth's axis in space cannot change appreciably during a catastrophic event.*

The same constraint therefore applies to the diurnal rotation and they take Velikovsky to task on his physics; he cited Indian, Babylonian, Mayan, Chinese and Egyptian memories of a former 360-day year. The authors prefer to say:

> *What would be modified is Earth's orbital semi-axis and period rather than the rotation frequency of the globe.*

In other words they suggest a change to the Earth's orbit rather than to the length of day.

Overall, computer simulations apart, the series of papers are not unlike many other studies that have been put out by adherents of Velikovsky and which usually receive the accolade of pseudo-science. It is only the mathematical modelling and the academic credentials of the authors that allow them to escape that categorisation. One has to doubt whether anyone lacking such status could achieve publication of Velikovsky-related theories on such prestigious academic platforms.

The above summary should serve as only an introduction, with apologies again for any omissions or abbreviation due to brevity; interested researchers should read the original papers in detail and form their own opinions.

Areas of General Agreement with the Z-hypothesis
As one would expect of a group of physicists, their physics is impeccable. Many enthusiasts and commentators on the subject of catastrophism, Atlantis, etc, may make loose references to pole-shifts and axis tilts without understanding that these are not the same thing; or that a vast external force must be applied to the Earth to bring about the latter. The co-authors do not make these errors. However, it is not clear whether the orbital mechanics used in their computer simulation have been peer reviewed.

The 2002 paper shows an illustration of the pole shift as the North Pole spirals away from its former position over Greenland to its present location. This shift, they propose, was the terminal-Pleistocene event. The suggested spiralling motion follows the mechanism of the *Chandler wobble*, but is not quite complete, as will be discussed below. Overall the computer model predicts a shift of the North Pole away from Greenland by perhaps as much as 17-18° — but leaving open a lesser magnitude. Note that this would still allow the South Pole to lie within East Antarctica.

The proposed close approach and tidal bulge, rather than an impact scenario, would explain the lack of any physical evidence of an impact at the end of the ice-age (There is no known 'crater'). It offers a mechanism that could change the geoid of the planet affecting rotational balance — thus triggering a pole shift — in the absence of hard evidence. However there remain other factors that it could not explain; both cited within the papers themselves and elsewhere.

The authors are clear in their rejection of a general rearrangement of the planetary orbits and conclude that Earth's orbit could be only marginally affected, as would the obliquity in space. The focus therefore falls upon *tidal forces* that could cause a *geographical* pole shift and the effect on the Pleistocene climate due to the *periodic returns* of Z and its dust veil.

Divergence from the Z-hypothesis

I shall try here to be constructive in analysis of the hypothesis and to avoid purely negative criticism.

Where the authors depart from their excellence in physics and astronomy then they stand on less firm ground. Their knowledge of climate and sea-level research is less well demonstrated in the papers. Another example would be their approach to Egyptology. One doubts whether any Egyptologist would consider a civilisation with written historical records, in Egypt, or Greece, as far back as 12,500 years ago. At that era the Nile valley was a swamp and the hunter-gatherers who would become Egyptians were able to live all over the still-green Sahara.

The Z hypothesis takes the literal dates supplied by Plato and Herodotus for the antiquity of Egyptian civilization and notes their close correspondence with the date of the Pleistocene-Holocene transition. This apparent correspondence of the terminal ice-age event and the pseudo-Egyptian chronology was a subject explored in the present author's earlier books. It can be explained by reference to the Egyptian belief, found in Manetho, that the antiquity of their state extended back 36,525 years with the dynasties of kings being preceded by a long list of demi-gods and god-kings. The Egyptians priests also made the same error as nineteenth-century Egyptologists in treating all the reigns as consecutive, whereas many kings ruled in parallel in different regions. The fact that we find the long chronology recorded in two independent sources dating from c.600 BC (Solon and Herodotus) shows us that this error is ancient.

The confusion in the position of sunrise and sunset is also explainable by standard Egyptology. The Egyptian civil calendar was not adjusted by an epagominal day (equivalent to Julio-

Gregorian 29 February) and therefore wandered through the seasons every 1,460 Julian years (*the Sothic Cycle*). Thus it sometimes had summer occurring in 'winter' months and vice-versa. This would indeed have cycled twice during the period of Dynastic Egypt from the predynastic up to the date of Herodotus. It therefore seems likely that there was a misunderstanding somewhere between the priests trying to explain this via an interpreter to their visitor, and in the understanding by Herodotus of what was being described to him. The simplest explanation is always the best and a simple mistranslation is far more likely than an astronomical cataclysm.

The problem of a former 360-day year is another fragile point in the Z-hypothesis. In the 2010 paper this world-wide calendar device is employed as a memory of the terminal-Pleistocene changes. However, the Indian and the Mayan eras, when these changes supposedly occurred, are placed at 3102 BC and 3114 BC in the respective mythologies. This would place it in the mid-Holocene rather than at the Pleistocene-Holocene transition. Moreover, it cannot be employed as a memory of a change to Earth's orbit when the conclusion from the authors' own simulation is that Earth's orbit was *'only marginally perturbed'*. Alternatively, a real modification to the length-of-day would require a significant change of angular momentum, which the authors agree, could not occur via this mechanism. The presence of a belief in a former 360-day year, in distant and unrelated ancient civilizations, must therefore remain an enigma that is not explained by this model.

The proposed pole shift itself is attributed to the mechanism of the *Chandler wobble*; being merely a much larger manifestation of the tiny pole-shifts that modern geophysicists regularly monitor. It is generally conceded that the cause of the modern wobble remains unknown and could therefore be a vestige of an ancient event. Current geophysical theory suggests that the mechanism would not be a neat spiral from former pole to new pole as shown in the authors' illustration. Rather, the pole must jump instantaneously to a new 'excitation pole' and the spiral wobble then commences; however as the proposed Z passes-by and the location of the bulge progresses, then the axis would jump repeatedly before the spiral completes. Only when the encounter was concluded would the wobble be allowed to decay to rest. Moreover, the Earth has a second mode of wobble, which should also be excited and would interfere with the Chandler wobble. A good mathematical explanation of the required excitation mechanisms is given in Lambeck's *Geophysical Geodesy* (pp 552-5). [4]

The co-authors say in the 2002/2018 paper: '*Note that this model has only few free parameters*'. This is an admission that: although it may be demonstrable by modelling that a stray-planet in just the right circumstances could cause the tidal effects (as they have defined them) it is all extremely unlikely. It has to be just right; in just the right orbit and just the right size; and it has to be just the right temperature to disintegrate and glow like a second sun. The hypothesis offers us a comforting view that this unwelcome visitor, which caused the ice ages, is now gone for ever and will not trouble us again. However, as the author's themselves conclude: "*On the other hand, it creates new problems that deserve a more complete treatment in future studies*".

The notion that we are now safe from a recurrence of the Ice Age is premature. We see during the Holocene a series of climate transitions and sea-level variations. There is a pattern of evidence across various disciplines that points to a lesser climate change and sea-level event in the mid-Holocene, during the late fourth millennium BC. In Europe, we also find transitions from one stable climate regime to another (the pollen zones) just as was ongoing during the Ice Age. The mid-Holocene: Atlantic to Sub-Boreal transition (known as Hypsithermal in North America) is the strongest of these signals. While fluctuations of temperature could be explained by a dust veil, rapid transitions from one stable regime to another cannot. These require further astronomical interventions similar to the terminal Pleistocene event. It is not clear how the fragments of Z, returning to our vicinity, could cause these unless they impacted; the fragments would not possess sufficient mass to pull a tidal bulge as is proposed for the main event.

A Mars-sized planet, at the distance of the Moon would appear six-times as large in our sky. If it must pass within 15,000 km of us in order to cause the required tidal effects then it would fill half the sky [See Note 4]. Furthermore, not only would it raise a solid-tide and an oceanic tidal bulge, but *surely*, it must also draw part of the atmosphere away into space? Noah, floating in his ark, would have no air to breath; and above his head the sky would be filled by a huge red planet. In other regions the ocean beds would be sucked dry and land at sea-level would be elevated temporarily into the stratosphere. Strangely no memory of this is described in the various Flood myths and religious texts from around the world; although we do find the myths of sky-gods, fiery dragons, second suns, etc; all of which could be explained by less drastic comet sightings. Commonsense would suggest that Planet Z did not happen, even though it remains a theoretical possibility that can be modelled in a computer.

A conundrum is that Z has to behave sometimes like a planet; sometimes like a comet. The authors admit that they have not modelled the change in mass as the repeated returns of Z cause it to break apart. It must retain the required mass to produce the solid tides; therefore, it must have a rocky, or even an iron core just like the real Mars. However, it also has to break-up and disappear after repeated encounters with both Earth and Sun. It should therefore have left a rocky asteroid belt, equivalent to that between Mars and Jupiter, rather than just a comet-like tail of dust and gas. Surely at least a few of these larger bodies should still be with us today? If Z is to be just a huge icy comet like, for example a larger version of 95P/Chiron, then it could not possess the required gravity.

Another consideration is the equatorial bulge. The polar radius is 6,356km while the equatorial radius is 6,378km; a difference of some 22km. For a postulated 18° pole shift then the pole tide generated is of the order of 6,500m. This is simply enormous – higher than the Alps. Nineteenth Century science discussed this and shied away. The Z model, so the co-authors argue, was based on a loose 1000 day period of relaxation within which the plastic-Earth would conform to the new ellipsoid. Geophysicists cannot yet give us any certainty as to how long a relaxation of figure might actually take in such hypothetical extreme circumstances. It should be apparent that during this transition the deep ocean would spill-over continental interiors, not just the coastal zones. It would sweep away fragile lake eco-systems such as Baikal, the survival of which the authors have already used to justify the former position of the poles. We should be finding the bones of Pleistocene whales in the desert.

Many of the objections to Z could be removed if it is not required to return repeatedly and if it is not employed as the principal cause of ice ages. The fundamental basis for ice-ages is the presence of a frozen continent at one or both of the poles, together with the Milankovitch cycles due to secular gravitational pull of the planets. These factors would be present regardless of additional interventions from space. A single close passage by a massive body, originating from the outer solar system or beyond, is more plausible. Also, the mass need not be just a planet or comet, if we allow the possibility of *mini-black-holes* recently proposed by Scholz and Unwin. [5] There may be more than one type of astronomical cause for past climate episodes and a Z-type body is just another to add to the list.

Causes of ice ages apart, as to whether the tidal bulge mechanism is the right one to explain the terminal ice-age event depends upon what it is required to explain. If it is to trigger solely a pole-shift then it remains a candidate; however if it must also explain a change of obliquity or the length of day, or a change to the orbit, then we must look for other ways to apply the necessary energy without leaving hard evidence.

In their closure summary (2007) the co-authors, as is usual, suggest the need for further study. In their own words: *'In the case that these contradict the given estimates, they could primarily question this particular scenario, rather than the evidence that a pole shift has occurred'*. So, we are in agreement that a pole shift occurred; the disagreement is: more than one and more than one possible cause.

Notes and References

Note 1
This author's own work on catastrophism, in parallel but slightly preceding the Z-hypothesis (published 1995-2005 and 2017) has always avoided comparisons with Velikovsky. [6] Since reading his books in the 1960s – long before I began any serious research into ancient catastrophism – I consciously avoided reading or citing them in order not to become associated with his unique astronomy – but there is no escape from it. The Z-hypothesis was just one more strand of Velikovsky-like research to evade. It would have been better if the authors of the Z-hypothesis had let it stand alone rather than citing Velikovsky in so many places, as I remain unconvinced that he actually believed everything that he wrote. In this author's books (1995-2017) a similar pole shift mechanism based on the Earth's wobble was explored primarily in support of a mid-Holocene event, but also acknowledged the occurrence of a much larger event at the Pleistocene-Holocene boundary. A discussion of alternative pole-shift mechanisms due to low-mass, high-energy impacts may be found at:
https://www.third-millennium.co.uk/dangers-from-ancient-supernovas
(or see Chapter 1 above)

Note 2
All comments here are based on the v1 versions of the subject papers available at *arxiv.org*. The older versions on the Los Alamos website (*lanl.gov*) that are cited within the papers themselves and elsewhere could not be obtained and have not been cited other than where the authors update their own thinking.

Note 3
An explanation of *Poynting-Robertson drag* may be found at:
https://academic.oup.com/mnras/article/460/1/802/2608912

Note 4
Earth's diameter is 12,742 km and the moon is 384,400 km distance; (30.168 diameters) The Roche limit for the Earth-Moon system is approx: 18,470 km (1.45 diameters). The solid earth tide due to the pull of sun and moon is of the order of 20 cm, and can exceed 30 cm.

Principal Sources Reviewed

1999
W. Wölfli, W. Baltensperger (1999) A possible explanation for Earth's climatic changes in the past few million years. http://xxx.lanl.gov/abs/physics/9907033 [*Link broken – see note 2 above*]

W. Wölfli, W. Baltensperger, A possible explanation for Earth's climatic changes in the past few million years, Notas de Fisica, Centro Brasileiro de Pesquisas Fisicas, Rio de Janeiro, CBPF-NF-031/99, (June 1999). More recent version: http://arxiv.org/abs/astro-ph/9909464v1 [*this link and the 2018 link go to the same page, where a 2018 pdf is shown*]

2002
W. Woelfli and W. Baltensperger (2002) An additional planet as a model for the Pleistocene Ice Age, CBPF, Notas de Fısica 007/02 or http://arxiv.org/abs/physics/0204004; http://arxiv.org/abs/physics/0204004v1

2006
W. Woelfli and W. Baltensperger (2006) Arctic East Siberia had a lower latitude in the Pleistocene, http://arxiv.org/abs/physics/0604029 or CBPF, Notas de F´ısica, NF-008/06, March 2006.

2007
W. Woelfli and W. Baltensperger (2007) On the change of latitude of Arctic East Siberia at the end of the Pleistocene, http://arxiv.org/abs/0704.2489

2010
W. Woelfli and W. Baltensperger (2010 September 26) Traditions connected with the pole shift model of the Pleistocene http://arxiv.org/abs/1009.5078v1

2018
R. Nufer, W. Baltensperger and W. Woelfli (2018 March 29) Long term behaviour of a hypothetical planet in a highly eccentric orbit. http://arxiv.org/abs/astro-ph/9909464; http://arxiv.org/abs/astro-ph/9909464v1

Other References

1. https://robertnufer.ch/

2. Palmer, Trevor (2018) *Perilous Planet Earth Revisited*, Chronology and Catastrophism Review 2018:2, 3-19
https://www.academia.edu/41250309/Perilous_Planet_Earth_Revisited_Chronology_and_Catastrophism_Review_2018_2_pp._3-19

3. Warlow, Peter (1982) *The Reversing Earth* ISBN 10: 0460044788 - ISBN 13: 9780460044783 - Weidenfeld & Nicolson, London.

4. Lambeck, Kurt (1988) *Geophysical Geodesy*: *The Slow Deformations of the Earth*. Clarendon Press, Oxford ISBN: 0-19-854438-3.

5. Scholz, Jakob and Unwin, James (2019). *What if Planet 9 is a primordial Black Hole?*
https://arxiv.org/pdf/1909.11090.pdf

6. Dunbavin, Paul (2005) *Under Ancient Skies, Ancient Astronomy and Terrestrial Catastrophism*, Third Millennium Publishing, Nottingham.
ISBN: 0-9525029-2-5 www.third-millennium.co.uk

5   The Astronomical Significance of Spirals
In Neolithic Art

Since the 1960's archaeologists have somewhat grudgingly accepted that many solstice and equinox alignments are evident at Neolithic stone circles and temples. We also find spiral decoration and motifs on some of these monuments and on other artefacts worldwide; most of these date from the third millennium BC (or between about 3100 BC and 1500 BC).

An extensive summary of these spirals and related cup-and-ring art in Britain and Ireland is given by van Hoek. [Note1] He suggests that many more Neolithic spirals have been misidentified as cup-and-ring marks because they are so badly worn.

In my own earlier books *Atlantis of the West* and *Under Ancient Skies* I proposed that this spiral art records a wobble of the Earth's axis of rotation, consequent upon an astronomical event that occurred in the late fourth millennium BC (around 3100 BC). In order for the Earth's axial tilt (the obliquity) to be modified requires an external force such as an energetic impact event to excite it, or perhaps something more exotic. It would then have to wobble until stability could again be restored.

The Earth has two modes of transient wobble, one (the Chandler Wobble) is damped after about twenty years, but the other mode is much longer-lived and may persist for perhaps two-and-a-half thousand years. It is vanishingly small on the Earth today and geophysicists still don't really understand what it might look like (should something trigger it). For a long-time the motion didn't even have a name (or rather it was misnamed) but geophysicists now term it the Free Core Nutation. It would occur if the principal axes of the Earth's liquid core and the outer mantle somehow became misaligned. [Note 2]

If a change of obliquity were in progress then it would become evident as a spiralling motion of the rotational pole about the celestial pole, causing seven-year rhythms in the weather and climate. However, it may not always be a nice neat spiral! Such systems can become chaotic, such as occurs with a double pendulum! [Note 3] Misaligned axes could present a similar problem. So one may see why ancient people, living under such an uncertain climate regime, would need to track the irregular spiralling motion of the sky in order to predict the seasons. It follows that ancient people would also have needed an accurate calendar against which to track an abnormal episode.

Here are a few examples of spiral art from the third millennium BC, with selected links, for anyone who wishes to research further into this phenomenon.

The Long Meg Standing Stone, Cumbria

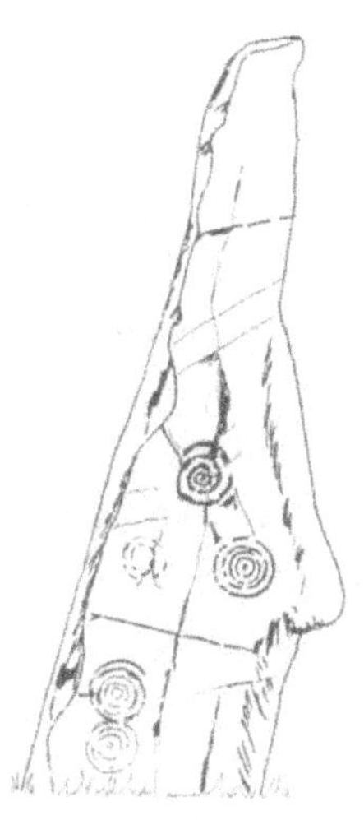

This drawing originally featured as fig.10.5 of *Atlantis of the West* and on the cover of: *The Atlantis Researches*. The notch marks mid-winter sunset on the western horizon as viewed from the stone circle of Long Meg and her Daughters. Date approximately 3000 BC. The drawing that I made in 1990 was art and not intended to be 100% accurate, but as you may see from the photographs, it is closer than some others you may see. The Long Meg stone circle remains much as it was built around five-thousand years ago.

The Long Meg Spiral and horizon views beyond

Mehen Spirals from Egypt

The spiral 'gaming board' discovered in the tomb of Hesy-Ra (Third Dynasty approx 3000BC) is usually associated with a game called Mehen, after the snake-deity. I suggested that this might be a 'spiral calendar' marking the height of the Nile flood over a seven-year cycle. It certainly looks like some form of game; but a spiral of seven turns is a remarkable coincidence. Another example is available in the British Museum and of course you can always find more details at: https://en.wikipedia.org/wiki/Mehen_(game)

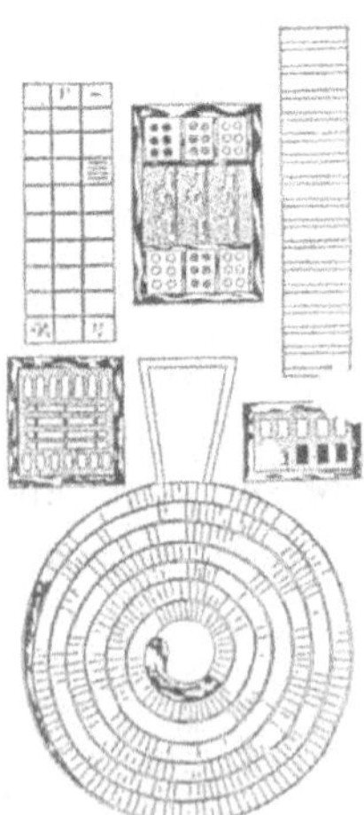

I am unaware of any spiral motifs on Egyptian monuments, however we do know that the Old Kingdom pyramid shafts were aligned on the celestial pole and prominent stars; and of course we have the climate references of the seven-good and seven-bad years of the Biblical Joseph story. I shall not open that particular box here!

The Phaistos Disk, Crete

Another spiral from this era is the double-sided Phaistos Disc, now in the Heraklion Museum, Crete. The Linear-A hieroglyphs remain undecipherable so we don't know if it was a calendar. It is loosely dated to the second millennium BC.

For some reason the publisher chose to use it as cover illustration for: *Atlantis of the West* – don't ask why! Authors have little say in such matters!

http://www.interkriti.org/crete/pg/?pg=1512101

The Spiral Stone, Isle of Man

An unspectacular monument at the roadside situated not far from Laxey and Cashtal-yn-Ard. The spirals may be seen at bottom left. Uncertain date, but probably the stone is Late Neolithic and not in its original location.

This spiral was not described in detail by van Hoek (although listed) but should rightly belong with his Galloway & Cumbria grouping. Alignment of the spirals resembles those at Long Meg, Cumbria.

https://www.waymarking.com/waymarks/WMY3PW_Neolithic_Spiral_Stone_Ballaragh_Isle_of_Man

The Calderstones, Liverpool

The recent history of this monument is a tragedy. The spirals are found on the six stones preserved in a greenhouse within Calderstones Park in Allerton. They were moved from their original location by the former owner during the nineteenth century, so we can no longer test any astronomical alignments. We can't even be sure whether they were from a stone circle, a chambered cairn, or part of an earlier court cairn dolmen. However, the decoration suggests they were contemporary with other decorated chambers in Ireland and Anglesey. The spirals themselves are incised and quite clear, unlike the Long Meg and Scottish examples; implying that they were mostly decorative art rather than practical instructions for observing the alignments. Pictures and history may be found in these links:

https://www.megalithic.co.uk/article.php?sid=6269
https://www.liverpoolpicturebook.com/2014/10/calderstonespark.html
https://www.themodernantiquarian.com/site/248/calderstones.html

Bryn Celli Ddu, Anglesey

An archeologically reconstructed mound that prevents us from proving any former astronomical alignment associated with the spiral. The 1847 engraving of the unreconstructed mound in the link below illustrates the problem! The spiral itself forms the head of an elaborate snake motif on the reconstructed entrance stone. Age is Late Neolithic. However, unlike other monuments discussed the alignment appears to have been towards the midsummer solstice rather than midwinter. The entire area shows evidence of later ritual use by the Druids right up to the period of the Roman destruction of their 'sacred groves'.

http://www.stone-circles.org.uk/stone/bryncelliddu.htm

For a discussion of 'roof boxes' at this and other passage graves see
http://www.stone-circles.org.uk/stone/bryncelliddu.htm

Newgrange, Ireland

The reconstructed 'passage grave' in the Boyne valley, Ireland exhibits spiral decoration on the external stones as well as a three-leafed spiral at the end of the main chamber. At midwinter sunrise a narrow beam of light from the 'roof box' above the entrance briefly illuminates this spiral motif. The beam therefore would have been ideal for tracking an abnormal variation of the sun's position as it moved along the walls. Probable date: 3150±100 BC from radiocarbon (see: O'Kelly 1974). Sceptics will suggest that the alignment at Newgrange is an artefact of the reconstruction. The nearby unreconstructed chambers at Knowth and Dowth offer a more reliable measure of ancient alignments.

https://www.nature.com/articles/337343a0
https://www.academia.edu/11482254/Computing_the_Winter_Solstice_At_N ewgrange_Was_Neolithic_Science_Equal_To_or_Better_Than_Ancient_Greek _or_Roman_Science
https://www.irishtimes.com/news/ireland/irish-news/newgrange-sun-trap-may-be-only-50-years-old-says-archaeologist-1.2913483

Images from the book by R.A.S. Macalister (1931)

A good summary of the legends about Newgrange and descriptions of the monument from before the site was disastrously reconstructed as a tourist attraction may be found at:

https://voicesfromthedawn.com/newgrange/

Knowth and Dowth, Ireland

The other passage graves of the Boyne Valley exhibit both spirals and alignments to midwinter sunrise and the equinox. Horizon alignments to the sun and moon could easily be obscured by bad weather, so it would make sense to use a range of seasonal alignments in order to increase the likelihood of good seeing conditions. Modern astronomers will know all about the frustrations of cloudy weather! A comprehensive (but for myself unconvincing) paper by Turler expands upon the earlier calendrical interpretations of Martin Brennan. These theories would treat the art as flattened-out representations of the helical spiralling of sun and moon in the sky as the year progresses. The principal reason why I doubt these explanations is that such complex observatories are not necessary to devise a practical calendar; Mayans, Indians, Chinese and Babylonians all succeeded in devising calendars without building passage graves or stone circles.

https://arxiv.org/ftp/arxiv/papers/1903/1903.07393.pdf
https://www.newgrange.com/knowth.htm
http://www.visual-arts-cork.com/prehistoric/knowth-megalithic-tomb.htm
http://www.megalithicireland.com/Dowth%20Passage%20Tomb.html
http://www.visual-arts-cork.com/prehistoric/knowth-megalithic-tomb.htm

The Scottish carved stone balls

These curiosities were found at various sites, of which the best example is the Towie Ball from Aberdeenshire, with its spiral carvings. They are often found associated with Late Neolithic stone circles and were carved with flint tools. An explanation of a practical application for such objects is awaited, as with the spiral 'calendars' discussed above.

https://edition.cnn.com/2018/06/29/world/scottish-carved-stone-balls/index.html
https://www.megalithic.co.uk/article.php?sid=2146412410

Tarxien and the Temples of Malta

The Maltese temples date from between 3500 BC and 2500 BC and are older than the Egyptian pyramids. The temple at Tarxien however dates from the later phase after 3000 BC. The spiral is clearly being used as decorative art, but again, the date is significant. According to author C.R Sant all the temples on Malta and Gozo exhibit both equinox and solstice alignments; and that all these alignments changed around 3000 BC. See his explanation below:

https://www2.stetson.edu/neolithic-studies/neolithic-art/neolithic-art-national-museum-of-archaeology-in-valletta-malta-2/
http://www.bradshawfoundation.com/malta/tarxien.php

According to Sant's reconstruction, the chambers of the oldest temples from the Ġgantija phase (3600–3200 BCE) are 'horseshoe' shaped calendars oriented towards equinox sunrise (an arrangement that would not be practical further north). At each solstice, the beams would fall upon stones beside the entrance of the 'horseshoe'. The solstice markers at Ggantija on Gozo formerly held spiral motifs less-ornate than those at Tarxien – however today these are almost unrecognisable. They show-up clearly on the watercolour paintings made by Charles De Brochtorff in 1829 when the temples were first crudely excavated. *See the paintings here:*

https://www.odysseyadventures.ca/articles/malta_temples/maltaTemples03e_ggantija.html

However, the Ta' Hagrat temple at Mgarr on Malta, is an anomaly among the Maltese temples focused towards the winter solstice sunrise (Sant p 54). This evolution parallels that found in Atlantic Europe, where the earlier court-cairn dolmens were (loosely) oriented east-west, perhaps towards the equinox, but gave way around 3200 BC to the solstice-aligned 'passage graves' and circles with spiral motifs.

https://www.megalithic.co.uk/article.php?sid=10424

Barclodiad y Gawres, Anglesey

Impressive as it looks, this mound is again a restoration by twentieth-century archaeologists (minus a roof-box). Frankly, I think that the 'honest' display of the Calderstones is better than this kind of false reconstruction; locked and inaccessible to casual visitors. However, there is no reason to doubt that the spirals themselves are authentic Neolithic examples. On the main stone are four spirals arranged in a row, flanked by another stone incised with a faint spiral that was missed by the original excavators. Other art on the stones reinforces the comparisons with passage graves in Ireland and Brittany. The link below provides excellent pictures.

http://www.stone-circles.org.uk/stone/barclodiadygawres.htm

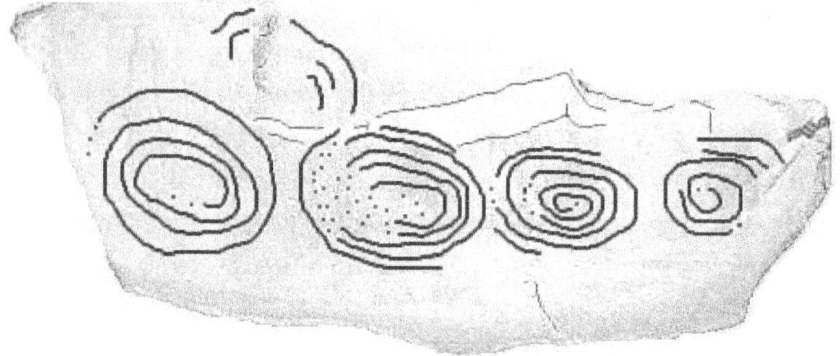

The author's crude drawing of the four faint spirals.

The drawing by *van Hoek* is not very accurate and I could not see some of the lines he depicted. However it seems clear that they are intended to be spirals.

Temple Wood, Argyll

Apparently an earlier Neolithic site that was demolished and the stones reused in the Late Neolithic to build a stone circle adjacent to it. A triple and a double spiral were described by van Hoek. Various alignments have been proposed, including one to midwinter sunset following the axis linking the centres of the old and new circles. I confess I could not find the spirals on the mossy stones, when I visited on a foul Scottish-weather day! They may have decorated a lost older monument and their modern position is chance – the Argyll coast is very rich in Neolithic sites.

https://www.thehazeltree.co.uk/2016/04/25/temple-wood-stone-circles-kilmartin-glen/
https://www.ancient-scotland.co.uk/site/143
https://www.themodernantiquarian.com/site/152/templewood.html

The Westray Stone, Pierowall, Westray

The most northerly example of a stone with decorative spiral art, discovered (and broken) by a digger in 1981 and presumed to be from a lost Neolithic site destroyed in antiquity. So again, nothing can be conjectured about its astronomical alignment. The art itself is reminiscent of the Irish Boyne passage graves and that of Brittany. We do find similar mounds in Orkney with aligned 'roof boxes', as at Maes Howe and the recently excavated chamber at Crantit, Orkney Mainland; the chamber faces south-east towards midwinter sunrise and appears to have lain undisturbed for 5,000 years. However, we cannot link these alignments directly to the spiral art. The Westray Stone is now in the island's Heritage Centre.

The Westray Stone (with pictures):
http://www.bbc.co.uk/ahistoryoftheworld/objects/pfbS6nxEQiOSvYEshpMOXQ
https://www.pinterest.co.uk/pin/1055599883016398/

Orkney passage graves:
http://www.orkneyjar.com/history/tombs/index.html
http://www.ancient-wisdom.com/scotlandcrantit.htm

Further examples of spirals may be found here...

Native American rock-art, Gavrinis, Gallicia and more...

Comment:

Regardless of the utility of the spiral to represent an astronomical phenomenon, it is always possible that the occurrence of any individual spiral motif, in other eras and contexts, could just be art; for example the Tarxien and Newgrange examples, as contrasted with those found inside the chambers or on aligned-stones. Context and dating is important. I would agree with van Hoek that cup-and-ring art may be older than spiral art, especially in a British-Irish context but disagree in that the use of spirals in so-called 'passage graves' should be older than those at outdoor stone circles. An internal beam is useful for practical measurement of where the sun is, whereas an outdoor alignment would be better for ritual purposes involving an audience. It is regrettable that in so many cases the original alignment associated with the spirals has been lost and cannot be tested, often exacerbated by modern archaeological reconstructions.

Note 1:

The study by van Hoek really cannot be praised enough. He concludes that the spiral art originated in south-west Scotland bordering the Irish Sea, based on the concentration of examples found there. I have tried not to unnecessarily repeat his work here, rather to complement it with additional British examples that he did not stress, together with other worldwide examples of spiral motifs on temples and artefacts.

Note 2:

In older geophysical papers the core-wobble was unrecognised and was termed the nearly-diurnal wobble, which is just the body-related part of the motion, whereas the core nutation is the component in space that Toomre (1974) reminded us should be some 460 times larger in amplitude. Due to the lack of a recognised name I used the name core-mantle precession in my book The Atlantis Researches: the Earth's Rotation in Mythology and Prehistory (1995). These names are all the same motion. For the non-geophysicists Toomre's paper is relatively lucid to read for a non-specialist seeking a 'plain-English' point of entry to this complex subject. [2] Generally speaking, geophysicists discuss theoretical motions in dense mathematics among themselves with little concern whether such theoretical excitations may actually have happened on the real Earth in recent prehistory; or that non-specialists might also have an interest in the subject. A more up to date bibliography of research is given in reference [5] and one of the clearest explanations of the Poinsot kinematics may be found on pages 52-58 of reference [6].

Note 3:
Would all please appreciate that everything discussed here is based on standard geophysics and takes no inspiration from Velikovsky! The core wobble of the Earth is a transient motion and could become chaotic in extreme circumstances as occurs with the double pendulum. See reference [3] below

References
1) van Hoek, Maarten A.M. (1993) The Spiral in British and Irish Neolithic Rock Art, Glasgow Archeological Journal, Vol 18, Issue 18
https://www.euppublishing.com/doi/pdfplus/10.3366/gas.1993.18.18.11

2) Toomre, A. (1974) On the nearly diurnal wobble of the earth. Geophysical Journal of the Royal Astronomical Society, 38(2):335–348, 1974. ISSN 1365-246X. doi: 10.1111/j.1365-246X.tb04126.x.
http://dx.doi.org/10.1111/j.1365-246X.1974.tb04126.x

3) https://www.youtube.com/watch?v=d0Z8wLLPNE0

4)
https://www.iers.org/IERS/EN/Science/EarthRotation/PolarMotionPlo t.html

5) Ferrándiz, José & Navarro, Juan & Escapa, Alberto & Getino, Juan. (2014). Earth's Rotation: A Challenging Problem in Mathematics and Physics. Pure and Applied Geophysics. 172. 57-74. 10.1007/s00024-014-0879-7.
https://www.researchgate.net/publication/271681663_Earth's_Rotati on_A_Challenging_Problem_in_Mathematics_and_Physics

6) Leick, Alfred (1978) The Observability of the Celestial Pole and its Nutations; Prepared for National Aeronautics and Space Administration, Goddard Space Flight Center, NSG 5265 jZ6 5; OSURF Project 711055, Reports of the Department of Geodetic Science, No. 262
https://ntrs.nasa.gov/search.jsp?R=19780025003

The preceding is a format-adapted version of the interactive webpage originally at:

https://www.third-millennium.co.uk/neolithic-spirals

6   Spirals on the Long Meg Standing Stone

The stone circle of Long Meg and her Daughters in Cumbria, northern England, is one of the easiest to reach as it lies not far from the M6 motorway; and it is also one of the most interesting because of the prominent spirals and cup-and-ring marks on the face of the Long Meg standing stone. In the 1990's I used it as an illustration of how such alignments might be used to track the position of the sun at the horizon in the event that the Earth's rotation were wobbling slightly, disturbing the seasons and agriculture at northerly latitudes. An article about the astronomical significance of spiral art at other Neolithic sites is available above with links to various surveys and opinions. [Ref: 1]

The Long Meg standing stone showing the western horizon behind
(click for links or find them in the list below)

It is generally accepted by archaeologists that the outlying Long Meg standing stone, is aligned on the midwinter sunset as viewed from the centre of the stone circle. Why choose midwinter? Quite simply, if the need is to forecast whether the spring and summer growing season will be blighted by unseasonal weather then you need to know before the season begins. Archaeologists now prefer that the alignments at Stonehenge and many other monuments were also directed towards the midwinter sunset rather than the midsummer sunrise that is ceremonially re-enacted each summer by the modern Druids.

Firstly: an aside by way of illustration. You may perhaps have noticed, as you sat to eat your breakfast one morning, how the rising sun shone directly into your eyes and you perhaps pulled the curtain over a little. Within a few days the sun will have moved round and the alignment is lost – but on the same day precisely a year later the identical annoying phenomenon will recur. You may not have realised it at the time but you were performing Neolithic horizon astronomy. It is actually quite easy to work out the number of days in the solar-year by setting-up a simple alignment like this to a celestial body and then count the days between recurrences; you don't need a complex Stonehenge-style circle, or a Newgrange passage-mound, in order to devise a calendar. The Babylonians, Indians, Egyptians and Mayans all created complex calendars without such devices. In the 1960's when Gerald S. Hawkins, Alexander Thom and others first proposed astronomical alignments at Stonehenge and other stone circles, they were initially ridiculed by the archaeologists of the day; as if the ancient Britons and Irish were not equally capable of performing simple naked-eye astronomy.

In my own books *Atlantis of the West* and *Under Ancient Skies* I proposed that the Earth's rotation was disturbed by an astronomical event in the late fourth-millennium BC. This would have triggered a spiralling motion due to a nutation of the Earth's axis in space. At an aligned monument such as Long Meg this would cause the points of rising and setting, of the sun, stars and planets, to migrate back-and-forth along the horizon over the period of the nutation. Therefore, I would not doubt the fieldwork on the subject of astronomical alignments at Neolithic sites – merely the stimulus that drove the ancient builders.

The free wobble of the Earth is a transient phenomenon; once triggered it decays exponentially to rest. The swinging of a pendulum is another type of transient. Another aside: suppose that I take you into a room with a stationary hanging pendulum then how could I prove to you that it was recently swinging and that the motion has since decayed to rest? How do you prove to me that it did not swing? Perhaps there was an eye-witness who recorded it – but would you believe the witness? Physicists can describe with equations the simple harmonic motion of a pendulum – but in certain circumstances the motion could become chaotic and unpredictable, as occurs with a double pendulum. A similar unpredictability could also apply to the Earth's rotation.

A pendulum would swing for ever unless the motion is damped. In the case of a pendulum it is air resistance and gravity that provides the damping; in the case of the Earth's rotation it is the elasticity and fluidity of the interior; properties that geophysicists have to determine from the tiny motions caused by earthquakes. At present the Earth's rotation is relatively stable and its nutations go unnoticed by us in everyday life. Geophysicists regularly measure the modern polar motion and the minor pole-shifts that have occurred since measurements began in the 1890s. Commonsense and limited imagination will lead us to assume that things have always been as they are now.

Perhaps the only way to prove whether the Earth's rotation was wobbling more significantly during the third millennium BC when the aligned monuments were built would be if someone or something recorded it. We have no historical texts from this remote era, only legends and the reports of later classical writers; but we do also have the astronomically aligned monuments with drawings of spirals and we also have natural climate records from such things as tree rings and ice cores; and pollen cores from bogs and lakes. Some of the underlying geophysics is explored below for those who wish to pursue the science a little deeper. [See Note 1] You should always read the small print!

The spiral markings on Long Meg are perhaps the most useful of all the examples of spiral-art found on Late Neolithic monuments. The circle has not been reconstructed so we may trust the alignments. On the stone itself we can discern a notch and a line (which appears man-made) from the edge to the spiral; and pointing towards the centre of the 'cup-and-ring' marks. Below it we see another notch and line pointing to a fainter example of a spiral with cup-and-rings; however the linking-line could be a natural fissure. Note also the parallel lines above the spirals and the straight line below. Unfortunately we see the monument after nearly five-thousand years of weathering and a modern wall now interferes with our horizon view beyond.

While we may trust the horizon alignment of the stones and the carving itself, we cannot be sure that the Long Meg stone has not leaned slightly over the years. By retro-calculation we can determine that the winter solstice sun around 3000 BC should have set at an inclination of around 22° to the horizon behind the stone

The upper and lower spiral carvings on Long Meg: the incised line seems to be pointing from the notch through the spiral towards the cup-and-rings that may denote the stone circle. The lower spiral and circle are less easy to see but in this instance the 'line' may be a natural fissure. There are also two faint concentric rings towards the left edge of the stone.

There is no point trying to be absolutely precise about retro-calculation as there are so many potential unknowns; but even allowing for some lean it seems unlikely that the inclination of the carved line has any astronomical significance. The spiral carving however seems to be telling us to observe the setting sun from the stone circle; as if it were supplying instructions for future generations as to how to use the monument.

As to whether the Long Meg stone has leaned-over since it was set in the ground then one may ask: what was the purpose of the straight lines and the two parallel lines? There is nothing artistic or decorative about them so clearly they had a practical purpose. The angle of the lower-line to horizontal is about 17° which we may compare with the inclination of the ecliptic (see retro-calculation above). If however, the stone were originally set with the lower line horizontal then the straight left edge with its notches would be approximately vertical. The parallel tracks are approximately at right-angles to the extended-line joining the spiral and rings; and the angle between this and the lower line is about 42°. Only excavation could determine whether the stone has leaned, or perhaps whether the markings survive from some previous use of the stone. Without such knowledge the various angles do not help us to understand the monument.

You may then ask: does any of this actually prove that the Earth's obliquity was disturbed during the Late Neolithic? No of course it doesn't. But pose the question the other way round. If the rotation axis were nodding and causing seasonal variations then how could you forecast it, if all you had were the tools available to Stone Age man? Answer: you would need an observatory something like a stone circle or the chamber of a 'passage grave'.

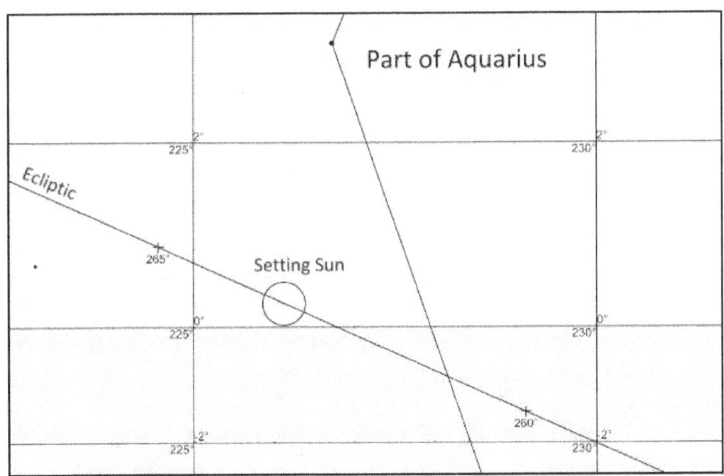

The stone circle of Long Meg and her Daughters (Google Maps)

Retro-calculation of the midwinter sunset in 3000 BC (Skymap Pro 6)

At northern latitudes the effect of abnormal seasonal variations would be more acute than at Mediterranean latitudes. The location of the Arctic Circle would fluctuate north and south of its modern latitude. It is the same as if you travelled north or south yourself to make the same horizon observations – except that it would be the pole that is moving not you. However this effect should not be overstated; a nutation of say, a degree or two of latitude would have noticeable effects on the climate, but no more than the equivalent of November or January weather conditions occurring in December (alternating suppressed and enhanced seasons). Such fluctuations would be a matter of survival for farmers at northerly latitudes, but in the physical record the evidence could easily be attributed to other factors by modern climate researchers.

Consider the following quotations on the subject of the arctic summer and winter, taken from Pliny's Natural History and from Julius Caesar's commentaries.

Pliny Book II LXXV. LXXVII
Thus it comes about that owing to the varied lengthening of daylight…parts of the earth have continuous days for 6 months at a time and continuous nights when the sun has withdrawn in the opposite direction towards winter. Pytheas of Marseilles writes that this occurs in the island of Thule, 6 days voyage N. from Britain and some declare it also to occur in the isle of Anglesey (Mona), which is about 200 miles from the British town of Colchester.

Julius Caesar V 13
Midway across is (Mona) and it is believed that there are also a number of smaller islands, in which according to some writers there is a month of perpetual darkness at the winter solstice. Our enquiries on this were always fruitless…

Other classical writers and sources could be cited, attesting to the competent astronomy of ancient people. Pliny wrote in the first century AD and Caesar was citing authors from the first century BC – before the Romans entirely destroyed the Anglesey Druids. Whatever ancient oral wisdom they preserved from antiquity was subsequently lost. Classical scholarship has always tended to disparage any link between the Druids and the astronomical alignments at Neolithic stone circles. This goes back to outmoded ideas about an invasion of Iron Age Celts from Gaul, who supposedly brought the Druids with them – even though the classical sources clearly tell us that the order originated in Britain. The links between Stonehenge and the Druids, so scholars told us, were all the invention of the eighteenth century antiquarian William Stukeley, citing Aubrey – and we still hear this dogma today, regularly trotted-out by the television academics. It is perfectly plausible that Iron Age druids and bards should have preserved ancient oral wisdom to be recorded in Roman times. So we do have eye-witness testimony recording the observations when the Earth wobbled on its axis. It's just a matter of whether you believe the witnesses.

As to whether a nutation of the axis could have carried the Arctic Circle quite as far south as Anglesey does seem doubtful. The Arctic Circle today, lies well north of the Shetlands and other references to *Thule* in Solinus and in Strabo's geography are clearly describing Lewis. If the Druids taught of an ancient time, when such extreme obliquity of the axis carried the Arctic Circle as far south as Anglesey then it might imply that the motion became chaotic in the immediate aftermath of the disturbance. However, it seems more likely that a lesser wobble of perhaps 1-3 degrees of latitude caused alarm at Orkney and Shetland, where we also find aligned passage graves and spiral motifs. Some archaeologists, following recent excavations, would like to regard Orkney as the Neolithic capital of Britain and that the building of passage graves and stone circles originated in the far north of Britain. [See Note 2]

Therefore it would be valuable if geophysicists would consider the evidence of the Neolithic monuments and the ancient literary and legendary sources; together with the 432 and 60 day cycles found in ancient Indian and Babylonian calendars. Perhaps then they could propose a coherent mathematical model (hopefully explained in simple-English) of how the real Earth would behave in extreme circumstances. Non-geophysicists such as archaeologists and historians could then cite this research and take it forward without being accused of speculation. This is the problem that one has when a transient motion has decayed to rest – how do you prove that it occurred? It is rather like ice on an aircraft's wings that has melted away before you can determine why it crashed.

Notes and References:

Note 1: The Small Print!

The present understanding of the Earth's free-wobbles has been hard-won by generations of geophysicists. The equations give two solutions; the first mode is the *Chandler Wobble*, named for its nineteenth-century discoverer; the second is the *core-wobble* or *free-core nutation* (sometimes called the *nearly-diurnal wobble*). To describe these motions without resort to equations and the specialist jargon is not at all easy. Both wobbles have a body and a space component as I will try to summarise in 'plain English' below together with some links for further research.

These 'free wobbles' should not be confused with the forced motions due to the constant pull of the Sun, Moon and planets on the equatorial bulge; these cause the phenomenon of the precession of the equinoxes and are an entirely separate consideration. Geophysicists usually refer to movements relative to a frame fixed in space as 'nutation' and reserve the term 'wobble' for the body-motion relative to a frame that rotates with the Earth. However usage of the terminology has been somewhat flexible in the past!

To recap the history of understanding, geophysicists first modelled the rotating Earth as a solid ellipsoid. This gives a period of 305 days for a 'solid' Earth as calculated by Euler. However when the true polar motion was first detected by Chandler the period was measured at 428-437 days (modern opinions vary). This was modelled by considering the Earth as an elastic shell containing fluid and by adding terms to Euler's equation for the elastic yielding of the shell and the fluid vorticity. Most models use variants of the Liouville Equation for which a good explanation may be found in references 2 & 4 below. No external force is required to act on the Earth; the modern Chandler wobble is excited by Earthquakes and deeper internal movements that slightly change the shape and balance of the planet. Using these as tools to probe the interior suggests that the time taken for the motion to decay exponentially to rest is around 68 years, but becoming insignificant after 20 years or so; this should be completely independent of its initial amplitude.

The free core nutation (FCN)) is more enigmatic. Geophysicists still argue about whether it has actually been observed; although it was known as a second solution to the equations since Hopkins in 1839 **and can only occur if the axes of the outer shell (the crust and mantle) Lm and the fluid core Lf became misaligned.** It should have a period of around 432 days according to IERS. [3] The second solution languished as a theoretical curiosity until the late twentieth century when some geophysicists claimed to have observed the nearly-diurnal wobble (NDFW). These are in fact two views of the same motion and the name led to much confusion. In my own earlier cross-disciplinary studies (in the absence of a then-recognised name) I referred to the spiral motion of the misaligned axes of core and mantle as *'core-mantle precession'*. The nutation and wobble should be visualised as occurring on top of this misalignment. On the Earth today, the FCN and NDFW are vanishingly small because the axes are aligned.

These various motions are not easy to visualise intuitively. The kinematic diagrams of Louis Poinsot may assist, for which one of the simplest explanations is found here on pages 52-58, in a paper originally intended for astronauts training. [4] It shows the conical representation describing the body and spatial components of the Earth's nutations. The point of the cones is at the centre of the Earth and the circles show the relative motion of the axis in space and within the body. Poinsot termed the path of the pole on the Earth's surface as the polhode; that of the pole in space as the herpolhode. The combined motion of the axis is represented by the line of contact of the two cones.

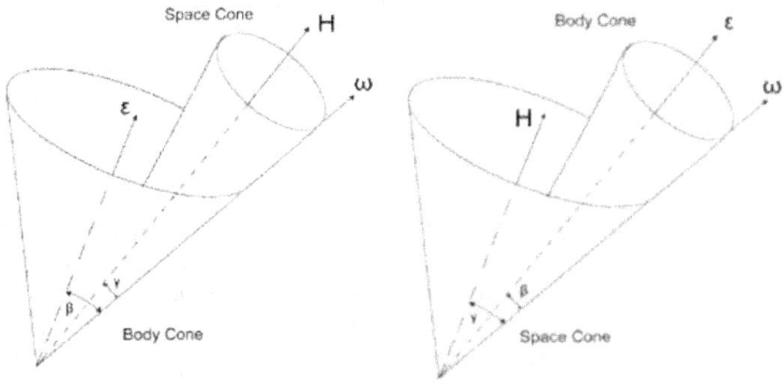

Poinsot diagrams for the Chandler wobble (left) and the core or nearly-diurnal wobble.

A useful illustration of the modern Chandler Wobble and its resultant pole shifts is provided by IERS. It may be seen that the rotation pole (the polhody) makes approximately six circuits of the celestial pole over a seven year period. This happens each time that the tiny modern motion is excited. It is important to note that the spiral motion illustrated is the polhody (the body wobble). For the modern Chandler Wobble the herpolhody (the motion of the axis in space) is so tiny on this scale that it would just be a dot in the centre of the diagram. However, for the core wobble it is the opposite. The Poinsot kinematics shows the much smaller body-cone rolling retrograde on the larger retrograde space-cone. The best estimates (guesses?) of the geophysicists as to how long the hypothetical motion would take to decay to rest suggest that it would persist much longer than the Chandler wobble; perhaps as much as 2500 years. The IERS suggested period of 432 days (other estimates say around 444 days) represents the theoretical 460 day period as clarified by Toomre, modified by the real characteristics of the Earth's interior. [See Note 3]

As Toomre cautiously put it in his landmark 1974 paper *"no process confined exclusively to either the core or the mantle can in principle bring about the misalignment of the angular momenta Lf and Lm that we have seen to be essential to this particular mode…Needed instead are random torques between the two major parts of the Earth"*. [5] This guarded phrasing implies an external force; something hitting or acting on the Earth from space, such as an energetic impact event. Geophysicists at that era could not openly discuss such topics, as to do so would have invited professional ridicule; hence in their research papers they speak cautiously about 'sources of excitation' and 'excitation functions' taking-place on purely theoretical Earth-models. It is not supposed to actually happen!

Therefore in order to consider a change of obliquity in ancient times (or an 'axis-tilt' in common parlance) then it implies an energetic impact event on the surface, or perhaps some other unknown force from outer space that could shake the core. It must excite both modes of wobble and for a time they would interfere, causing as Alar Toomre cautiously put it, the core motion to: "...*modify the basic rigid-body mode, the Chandler wobble, by perhaps one part in ten*". But after the Chandler wobble had decayed to rest, the core wobble would persist for many generations, thereby affecting seasonal weather patterns. The interference of the Chandler wobble during the first twenty years or so is again a separate consideration; it would not be a pleasant time to be on the Earth!

Implicit in all this guarded phraseology is that the so-called 'random torques', these being external forces, must change the angular momentum and obliquity and, since the shape of the planet is thus modified, it would also cause a pole-shift. While the core wobble is in progress nothing prevents the Chandler wobble from being excited again by purely internal changes to the figure of the Earth until full stability is restored.

Therefore should any non-geophysicist wish to propose a theory of pole-shifts or axis-tilts in the ancient past without having it dismissed by sceptics as Velikovsky pseudo-science; or linked with Dodwell's 'proof of the bible', or Hapgood's sliding-crust nonsense, then they must grapple with this geophysical terminology and the equations in the specialist papers. Similarly, the geophysicists are neglecting valuable data about the Earth's interior if they fail to consider the ancient astronomy.

Note 2:

The suggestion that Orkney was the 'Neolithic capital' of Britain was made in a BBC television program about the Ness of Brodgar excavations. As so often when a new theory about prehistory is suggested, the establishment archaeologists pounced on it. I hope my discussion above will perhaps offer some motivation behind the building of aligned monuments in the North, but of course it too will be immediately dismissed as is the academic norm with such things. Follow the links below and form your own opinion:

https://www.heraldscotland.com/news/15049513.bbc-historians-under-fire-for-claim-orkney-was-uk-capital/
https://www.bbc.co.uk/programmes/b08819tl https://www.nessofbrodgar.co.uk/2019/

Note 3:

The links here to the website of IERS (*International Earth Rotation Reference Systems Service*) will give access to further geophysical research on the nature of the Earth's core. There is now a special department researching the core: *The Special Bureau for the Core (SBC)*. Still largely unknown is the contribution of the solid nickel-iron inner-core; and the extent to which it would further modify any significant wobble and its effect at the surface. See Reference [3] below.

This article is written for the benefit of the general reader and is not intended to be an academic paper. The following references and the links above will lead to a comprehensive bibliography of relevant research.

References

1) https://www.third-millennium.co.uk/neolithic-spirals

2) Ferrándiz, José & Navarro, Juan & Escapa, Alberto & Getino, Juan. (2014). Earth's Rotation: A Challenging Problem in Mathematics and Physics. Pure and Applied Geophysics. 172. 57-74. 10.1007/s00024-014-0879-7.
https://www.researchgate.net/publication/271681663_Earth's_Rotation_A_Ch allenging_Problem_in_Mathematics_and_Physics

3) IERS: http://sbc.oma.be/freecornu.html and http://sbc.oma.be/

4) Leick, Alfred (1978) The Observability of the Celestial Pole and its Nutations; Prepared for National Aeronautics and Space Administration, Goddard Space Flight Center, NSG 5265 jZ6 5; OSURF Project 711055, Reports of the Department of Geodetic Science, No. 262
https://ntrs.nasa.gov/search.jsp?R=19780025003

5) Toomre, A. (1974) On the nearly diurnal wobble of the earth. Geophysical Journal of the Royal Astronomical Society, 38(2):335–348, 1974. ISSN 1365-246X. doi: 10.1111/j.1365-246X.tb04126.x.
http://dx.doi.org/10.1111/j.1365-246X.1974.tb04126.x

Hyperlinks

Long Meg Stone Circle
https://altogetherarchaeology.org/reports.php
A 2012 survey report on the Long Meg stone circle by Durham University downloads directly here.

https://altogetherarchaeology.org/Reports%20and%20Proposal%20Docs/Long %20Meg/AAmodule1cLongMeg2015excavationProject%20Design.pdf
http://www.stone-circles.org.uk/stone/longmeg.htm
https://www.youtube.com/watch?v=BHAxVwlWXf0
https://www.google.co.uk/maps/@54.7280261,-2.6675773,171m/data=!3m1!1e3

Simple Harmonic Motion
https://demonstrations.wolfram.com/ThePendulumFromSimpleHarmonicMotio nToChaos/

Double Pendulum
https://www.youtube.com/watch?v=d0Z8wLLPNE0

Chandler Wobble
https://www.iers.org/IERS/EN/Science/EarthRotation/PolarMotionPlot.html
https://en.wikipedia.org/wiki/Chandler_wobble

Liouville Equation
http://adsabs.harvard.edu/full/1966SAOSR.236..193C

Free core nutation and the nearly-diurnal wobble
http://sbc.oma.be/freecornu.html
https://academic.oup.com/gji/article/38/2/349/607495

Polhode and Herpalhode
https://en.wikipedia.org/wiki/Polhode
https://en.wikipedia.org/wiki/Herpolhode

This article is a format-adapted version of that available on the author's website at:

https://www.third-millennium.co.uk/spirals-on-long-meg

7   Updating George F. Dodwell

Summary: *In his retirement up to 1959 the Australian astronomer George Frederick Dodwell conducted extensive research to check whether the standard formula for the secular variation of the obliquity of the ecliptic could be proven by ancient observations of the sun at the solstices. He came to a conclusion that the formula could not be proven; and that in fact the data showed the likelihood of a catastrophe around 2345 BC, which he equated with the Flood of Noah. His work was never published, but it continues to be cited in all sorts of contexts despite being based upon science that is now (2005) half a century old. This article will examine how well his conclusions stand-up after fifty years of new discoveries.*

George Frederick Dodwell (1879-1963) was the Government astronomer for South Australia for forty-three years until his retirement in 1952; responsible among many notable achievements, for the accurate determination of Australian state boundaries. In his later years he seems to have pursued a strong personal interest in the secular variation of the obliquity of the ecliptic.[1]

That Dodwell also held a profound, yet scientific, interest in the Bible cannot be doubted, for in 1959 he completed his manuscript entitled: *The Truth of the Bible: astronomical investigations of the obliquity of the Ecliptic.*[2] The work in two volumes, comprising some 327 pages, was unpublished - yet it remains a milestone in the history of astronomy.[3] His manuscript does offer a unique mix of science and faith, but any religious inspiration for the research must not be overstressed. His approach remains scientific throughout.

The manuscript is structured with an introduction and summary chapter, where he presents his conclusions and charts. Chapter 2 considers the likely errors in the ancient data, with a discussion of the accuracy of the various types of gnomon instruments that were used to measure the solar shadow. He then offers a detailed discussion of each source, taking care to adjust all his results for refraction, solar parallax and reduction to the centre of the sun's disc:

Chapter 3 – Ancient Chinese
Chapter 4 – Ancient Hindu
Chapter 5 – Greek sources
Chapter 6 – Medieval Arab sources
Chapter 7 – Medieval and modern sources
Chapter 8 – Ancient Egyptian Monuments

Chapters 9 and 10 respectively consider the Stonehenge alignments and those of Tiahuanaco in Peru, although he finds no datable measurements from these monuments and makes no use of them in his calculations.

It is clear from the summary that Dodwell believed a catastrophic flood of the sea, akin to that recorded in the Bible, must be accompanied by a tilt of the Earth's axis. It is less certain however, whether he originally set out to prove the authenticity of the Flood by science, or whether it was his study of the secular variation which subsequently led him to believe that it showed a catastrophic element in recent Earth history.[4]

Dodwell's work owes much of its recognition beyond Australia to a letter that he wrote to the author Rene Norbergen in 1960, which was quoted in Norbergen's highly successful book *Secrets of the Lost Races*. The book details that author's expedition to Mount Ararat; and his claim to have discovered remains of Noah's Ark had evidently caught Dodwell's attention. Norbergen quotes from Dodwell's letter:

I have been making during the last 26 years an extensive investigation of what we know in astronomy as the secular variation of the obliquity of the ecliptic, and from a story of the available ancient observations of the position of the sun at the solstices during the last three thousand years, I find a curve, which after allowing for all known changes, shows a typical exponential curve of recovery of the earth's axis after a sudden change from a former nearly vertical position to an inclination of 26½ degrees, during the interval of the succeeding 3,194 years to AD 1850...The date of the change in the earth's axis, 2345 BC...

Dodwell concluded that this event was none other than the Biblical Flood and that the story of Noah must therefore be historically true.[5]

Therefore in 1998, while researching the astronomy for my own later book *Under Ancient Skies,* I attempted to look further into Dodwell's work on the obliquity[6] Unfortunately, I was forbidden by Dodwell's heirs to quote from the manuscript;** neither would they allow me to make a copy for the purpose of checking his source material in more detail.[7]

A trawl of internet references will bring up numerous citations by religious groups and creationists, each seeking to utilise Dodwell's professional credentials as proof of the Bible.[8] Witness a typical example taken from a (2005) creationist website: [9]

After further calculations which he [Dodwell] plotted on a graph it showed it was a log-sins curve showing that the spinning earth returned to a new angle when it had been suddenly deflected from its original position. He said it had happened around 2,345 BC. This would have happened from the start of the flood (2,470) and lasted many years until the birth of Peleg (c2,372) when the continents were formed. Dodwell later read Professor McReady Price's book; "Hyphothesis of a World Catastrophe" and he realised that he had discovered the cause of the Noahian flood!

Dodwell's manuscript was also cited by M. M. Mandelkehr in his various articles in the C & C Review;[10] and in his recent books, to support a case for a disturbance of the earth's axis around 2300 BC,[11] This date would loosely coincide with the First Intermediate Period of Egyptian history, as it is conventionally dated. Various authors from the 1970s and early 1980s also made a case for an astronomical catastrophe around this era, of which perhaps the most influential examples would be Clube and Napier's book *The Cosmic Serpent*.[12]

Many academic papers up to the mid-1970's published *uncalibrated* radiocarbon dates for climate and sea-level changes c.2300 BC which, when calibrated, have to be pushed back by around 800 years. Once this is done their apparent correspondence with the Biblical dating of Noah's Flood is lost.[13] Despite this, creationists and others continue to cite Dodwell and other older sources as scientific support for the Biblical chronology.

In 1981 a discussion of Dodwell's conclusions was included in the SIS Journal, within a *Focus* article entitled *Catastrophism Old and New.* It minutes the presentation of speaker's Peter Warlow and Peter James at the Society's meeting of 6 June 1981.[14] The diagram included here, showing a simplified version of Dodwell's graph, is reproduced from a 1978 article entitled: *The Celestial Dynamics of "Worlds in Collision"*.[15] It captures the essence of, but differs significantly from Dodwell's original curve.

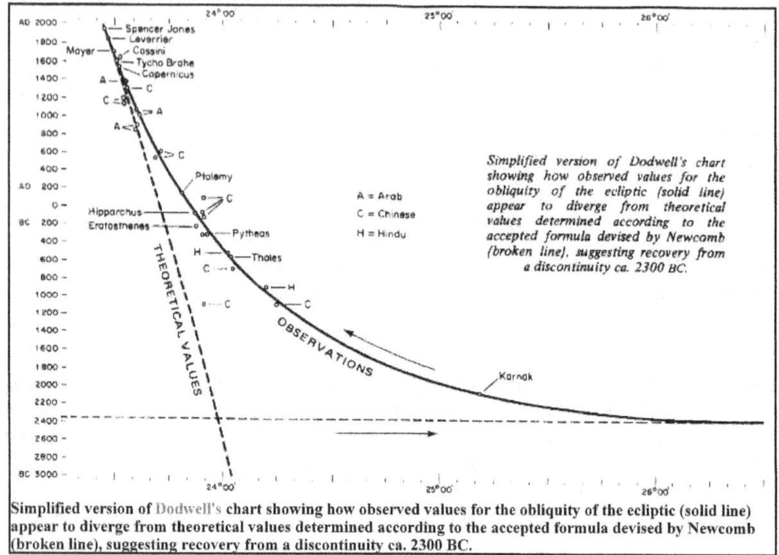

Simplified version of Dodwell's chart showing how observed values for the obliquity of the ecliptic (solid line) appear to diverge from theoretical values determined according to the accepted formula devised by Newcomb (broken line), suggesting recovery from a discontinuity ca. 2300 BC.

Reproduced by courtesy of the Society for Interdisciplinary Studies

The graph plots an exponential decay curve passing through a scatter of observations of the sun at the solstices, as derived from modern, medieval and classical sources, plus earlier Chinese and Indian observations from the first millennium BC. The curve then continues *precisely* through two older alignments derived from the Egyptian temple of Karnak, to an asymptote at 2345 BC, which Dodwell equated with the date of Noah's Flood.

Some significant additional details appear on Dodwell's original graph. He believed that the curve of adjustment of the axis had commenced at 2345 BC (he was most precise in his comparison with Archbishop Ussher's date for the Biblical flood) but had completely decayed by 1850. His summary diagrams are presented in portrait rather than landscape, with many more recent observations crammed into the curve. He actually shows *two* observations from Karnak, one dated to 1570 BC, which he admits is an interpolated value; and the earliest at 2045 BC, which he attributes to the Twelfth Dynasty and states the obliquity as 25° 09′ 55″.[16] The dotted line of 'theoretical values' were derived from Newcomb's formula.[17] This gives the uniformitarian view of the 'secular' variation of the obliquity of the ecliptic that most astronomers had long accepted; and which Dodwell set out to check against contemporary observations.

The 1978 article continues with a minute of a discussion session, in which a speaker, Dr Robert Bass, introduced the subject of Dodwell's curve. If I may quote him here:

When he [Dodwell] *charted these historical observations (see figure), they led to a curve very different from that given by Newcomb's formula, which most people think would be valid for thousands of years. His interpretation of his data indicated that the earth is recovering from being turned upside down around 2300 BC...*

This shows us how unintentional bias may creep-in through second-hand reporting, in this case when it was cited to support Velikovsky's theories. As we may see from the letter quoted above, this certainly was *not* Dodwell's own opinion. He did not state that his data proves the Earth turned upside down around 2300 BC; quite the contrary. Dodwell thought that the obliquity of the Earth's axis had been closer to vertical before the crisis (perhaps only $5°$) – although on what evidence he believed it to have been near-vertical is equally unsubstantiated. The references that he intended to cite are left as empty brackets and ellipses in the manuscript.

For let us be sure of one thing: the recovery of a rotating body, following a disturbance, can tell us nothing of the attitude of its axis *before* that event; to establish that, we must consult other data. We can only assert that *after* a disturbance of the axis (and barring any subsequent agitation) it would wobble in the manner established by geophysical theory, before ultimately settling to its modern attitude.

It is evident that Dodwell also knew something of contemporary geophysical theory, for he considers his own curve to be the result of an "Eulerian nutation"; and although he refers to the modified 430-day period of the Chandler Wobble rather than Euler's 305-day period, he never refers to it by that name. He also presents a modified curve, showing the effect of an oscillation of period 600 years, caused, he says by latitude variation *"accounted for by the inertia of the rotating body alternately retarding and hurrying the precession"*.[18]

Another fact about which we must be clear is that, despite his faith in the Biblical account of the Flood, George Frederick Dodwell was in every way a conventional professional astronomer of the early twentieth century. His reaction to Velikovsky's unique form of catastrophist astronomy would surely have been as negative as that of his professional contemporaries. We must however, view his conclusions in the light of the astronomical and physical science of the 1950's, for much of that background is now outmoded.

We may see that for the modern and medieval period back to 400 AD, Dodwell derives some sixty dateable values for the obliquity, but these become progressively fewer as we go back in time. For the early centuries BC he utilises points derived from Ptolemy, Hipparchus, Pytheas and Thales as the oldest of the classical sources. He plots the curve to pass through Thales observation at 558 BC leaving the other points as a scatter around the curve; Dowell also considered his postulated 'oscillations' to give a more precise fit to the scatter of points, although this is distinct only in the medieval and modern period where most of his data points lie. The curve then continues *precisely* through his two oldest points derived from the Egyptian temple of Karnak at 1570 BC and 2045 BC.[19]

The two early Egyptian data points, taken alone, are insufficient evidence to plot such a graph; ideally we need to see a statistical scatter of many more ancient observations clustered around this early part of the curve. If these two vital points should be considered unreliable then we can as well reconstruct the curve to pass through almost any date we wish.

The first problem arises with modern dating for Dynasty XII as lying between 1991 BC and 1786 BC.[20] The oldest phase of Karnak: of which only the granite thresholds survive, is believed to have been built by Amenemhat I, the first king of the dynasty. This would place Dodwell's oldest point at c1991 BC not at 2045 BC. However there is some reason to believe that this pharaoh only restored an older Amun temple built by a Dynasty XI king Intef II, who ruled this Nome during the First Intermediate Period.

In the 1950's, conventional Egyptologists held that the Karnak solstice alignment was mere coincidence. The pioneering work of J. Norman Lockyer, had originally suggested that the main axis was aligned towards the midsummer sunset (or rather toward clock stars that heralded this event at various epochs).[21] However, interest in the earliest alignments was renewed by Gerald Hawkins in the 1970's.[22] Based upon some difficult inscriptions within the temple itself, most Egyptologists now accept that the alignment was actually in the opposite direction towards the winter solstice sunrise. The surviving temple preserves the last attempt by a native king: Nectanebo I (380-362 BC) to restore the old monuments. The sun would have risen behind the Gate of Nectanebo, being framed within a rectangular slot high up in the gate.[23] We may therefore reasonably infer that the earlier phases of the temple were similarly oriented towards midwinter sunrise.[24]

The observations of Thales (believed to date from 558 BC) are another crucial point on Dodwell's curve, yet precisely how Thales performed his determination of 24°, which Dodwell adjusts to 24° 00' 56", must remain uncertain. Diogenes Laertius (I.24) recorded that Thales was the first astronomer to determine the true length of the year and the sun's course from solstice to solstice; and that he acknowledged the lost astronomy of Eudemus as his primary source.

The various fragments of Pytheas, speak of the Arctic Circle lying some six days sailing north of Britain, from which only loose information may be derived about the solstice. Dodwell derives a value of 23° 53' 46" for the obliquity at 323 BC, based on the latitude of Marseilles from where Pytheas set out in search of Britain and Thule. However, we also know that Pytheas considered Marseilles and Byzantium (Istanbul) to lie on the same parallel, although these differ in latitude by as much as one-and-a-half full degrees.[25] This shows us the degree of approximation that was sufficient for his navigation.

The observations of Eratosthenes (probably dating from 205 BC) lie slightly off Dodwell's curve. Eratosthenes' method for determining the size of the Earth relied upon an observation that on the day of the summer solstice, the sun penetrated precisely to the bottom of a deep well at Syene south of Thebes in Egypt. Yet Syene lies some 50 km north of today's Tropic of Cancer. According to Ptolemy, Eratosthenes derived the tilt of the axis as 11/83 of 180°, plotted as 23° 51' 59" by Dodwell. The graph of 'theoretical values' shows a retro-calculated angle of approximately 23° 43' 12.2" at 200 BC. However, consider how tiny this difference is in comparison to the various observational uncertainties. We cannot know the depth of the water in the well or whether its walls were vertical; indeed we cannot be sure whether the observation was a measured angle using a gnomon or just an anecdotal report.[26]

Older measures of obliquity are available from Indian and Chinese documents. Dodwell's oldest Hindu source dates from around 900 BC (in another place he says 945 BC) where he uses a value of 24° 11' 04" based on observations in southern Sri Lanka; but again the date and location can only be known approximately.[27] Chinese documents supply observations around 700 BC and 1120 BC. The corrected obliquity for Mo in Shantung province gives 24° 12' 06" for 1120 BC.[28] Dodwell noted that all these derivations were trending consistently above the prediction of Newcomb's formula.

When Dodwell was gathering his data during the 1950's, radiocarbon dating was in its infancy and the technique of tree-ring calibration was yet to come. We can now examine other dateable solstice alignments that were unavailable to him. The archaeological textbooks of that decade gave the era of Stonehenge and other aligned Neolithic monuments as no older than 1800 BC based on cross-dating techniques. It is now generally accepted that the first phase at Stonehenge – though possibly aligned towards midwinter sunset rather than midsummer sunrise – must date to as early as 2900 - 3000 BC. This shows us that the alignment of the solstices were certainly close to their present state fully six centuries before the date that Dodwell derived for the Biblical Flood. A change in the tilt would affect the azimuth of the rising or setting sun at the solstices; and better astronomers than I have concluded that the early Neolithic monuments do indeed agree with retro-calculated alignments; Gerald Hawkins, Alexander Thom and others would never have been able to make their case for Neolithic alignments if that were not so!

For example, 1960's excavations at Newgrange in Ireland and contemporary passage graves in Anglesey and Orkney show that these monuments were also aligned towards the solstices. It is well established that at midwinter sunrise, a beam of light penetrates a narrow slot known as a roof box to illuminate the back wall of the Newgrange chamber. A survey by the excavators established that the winter solstice sunrise would still penetrate the chamber regardless of secular changes in the obliquity of the ecliptic.[29] The Newgrange mound was radiocarbon dated to a date around 3150±100 BC from charcoal remains found between the stones.[30] This sets a limit to the most recent era at which any flood catastrophe could have occurred; and also indicates the range and variation of the obliquity at that era.

Perhaps the most significant advance since the 1950's has been in the geophysics and the understanding of the Earth's wobble. Astronomers of that era certainly believed that they understood this phenomenon, but experience has proven otherwise. It had been understood since as long ago as 1839 that a disturbance of the rotation axis would give rise to two modes of free wobble. One has a short lifetime, decaying exponentially within about 20 years (the *Chandler Wobble*, named after its nineteenth century discoverer). The other mode has a longer lifetime (between 2,000 and 5,000 years has been suggested) and it can only be the exponential decay of *this* wobble to which George Dodwell was alluding in his research. It would occur should the axes of mantle and core become misaligned.

It was not until the 1980's that geophysicists actually proved the existence of this theoretical motion.[31] Although it is vanishingly small on the modern Earth, the data shows that it causes a wobble of the axis in space of period approximately 430-440 days, together with a body-motion component, known as the *nearly-diurnal wobble* on account of its retrograde period of just under one day.[32]

The true characteristics of this core-wobble, as a cause of latitude variations, were unknown to astronomers of the 1950's. Indeed the subject was of interest to only a few of the most eminent geophysicists. Geophysical textbooks such as that of Sir Harold Jeffreys, the foremost authority of the time, were available but apparently not used by Dodwell.[33]

The axis cannot simply tilt over gradually as Dodwell's graph implies, for that would require the period of the wobble (i.e. each circuit of the pole) to be precisely one year, in order for the horizon alignments to return close to the same rising and setting each annual solstice. However, since the true period is closer to 440 days it would combine with the annual 365-day period to give a seven-year rhythm.[34] The solstice alignment would therefore lie sometimes north, sometimes south of its mean position (albeit following an exponential decay curve). This mean position, to which the motion would ultimately decay, is given by the modern rising and setting points after suitable adjustment for the secular variation at each era.

Dodwell's original calculation of the secular variation was based on Newcomb's 1894 and 1906 formulae. Newcomb had only intended his approximations to be accurate for 250 years either side of an 1850 epoch. His calculation for the pull due to Saturn was slightly incorrect and he did not know of the small effect of Pluto. Other astronomers from the 1950's onwards also began to question Newcomb's equation. So in 1976 (based on computer-calculated analyses of ancient and medieval data similar to that used by Dodwell) the International Astronomical Union adapted the revised formulae of Lieske.[35] These astronomical constants continue to be refined and extended.[36]

Even if we had many more ancient observations marking the extreme limits of a wobble at any particular era, then we should expect the resulting graph to show plots both *above* and *below* the theoretical line given by the retro-calculated secular variation. To use Dodwell's own statement: if the maximum of the disturbed obliquity were indeed some 26.5° then this lies some 3° above the mean; so we should also expect to find some evidence for points lying between a *minimum* obliquity of 20.5° and the mean of 23.5°.

However it would make no sense for ancient builders to align a solar temple on either of these extremes, for it would become immediately obsolete as the sun would never again reach it. It would be far better to align upon the mid-point of the wobble, through which it would pass each 3 or 4 years and would ultimately converge. The absence of any points on Dodwell's graph showing an axis tilt *less* than that of the present day, in addition to the points of increased obliquity, shows us that the evidence is incomplete. Alignments to the mid-point of a transient motion can neither prove nor disprove its former existence! For example: release a pendulum and wait for it to come to rest; how do you prove that it ever swung at all?

In summary it must be concluded that much as a catastrophist researcher might wish Dodwell's work to support a case that the Earth's axis has changed, unfortunately it cannot reliably be used in that way. Certainly it cannot be used to support the Biblical chronology as he intended. However, it may offer evidence that the axis was indeed disturbed at some point in antiquity; and that a residual free nutation remained detectable up to medieval times.

Nothing in my conclusions here should be taken as a criticism of Dodwell, for it remains a remarkable insight for an astronomer of his day to even consider the possibility of a change of axis, let alone to attempt to prove it. It is a tragedy that Dodwell's research did not achieve publication in the 1960's. It might have generated serious interest in catastrophist astronomy among professional astronomers; whereas the fringe debate, dominated by the ideas of Velikovsky – another piece of outmoded 1950's science – has simply caused them to look the other way.

****

****Note:** when this paper was written in 2005 the family were still not allowing publication of any part of the manuscript, but it is now freely available on the internet, including Dodwell's original curve as is reproduced here.. The trail of how I obtained a copy in 2003 is detailed in my own unpublished paper (above). Suffice to say that the ultimate source was Barry Setterfield, who later put it on his own website in 2010 – with a plea from the family not to quote it out-of-context! The manuscript should now be considered of mainly historiographical interest.

http://www.setterfield.org/Dodwell/Dodwell_Manuscript_1.html

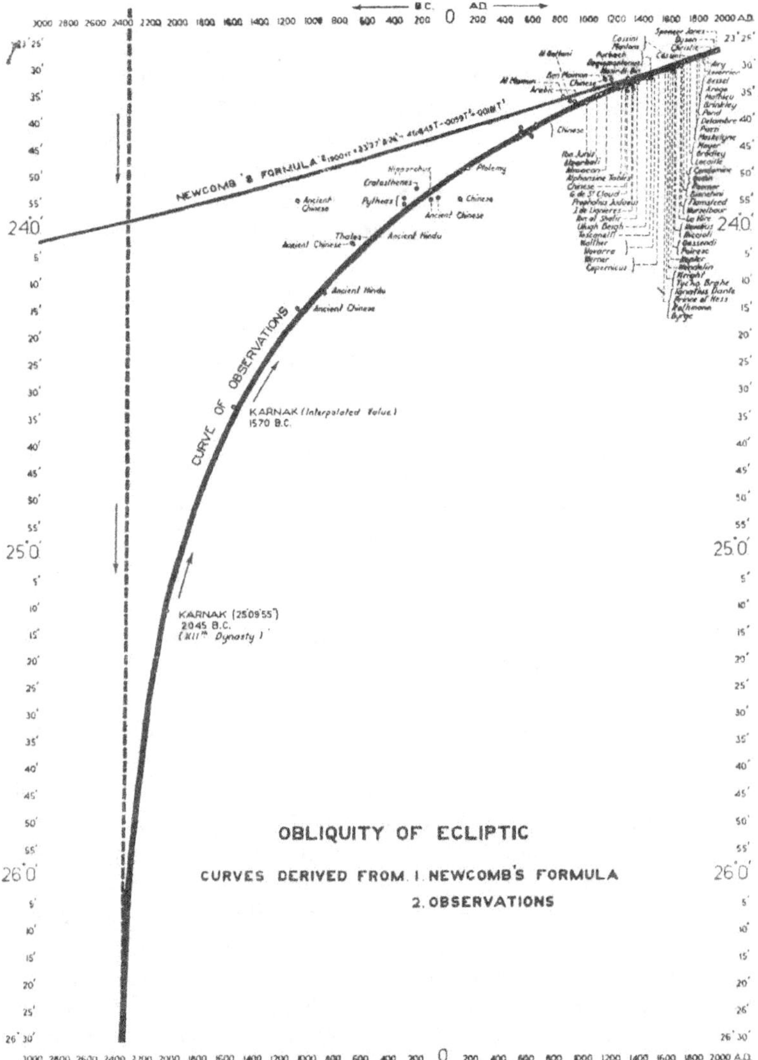

B.C. A.D.
3000 2800 2600 2400 2200 2000 1800 1600 1400 1200 1000 800 600 400 200 0 200 400 600 800 1000 1200 1400 1600 1800 2000 A.D.
NEWCOMB'S FORMULA
CURVE OF OBSERVATIONS
KARNAK (Interpolated Value) 1570 B.C.
KARNAK (25°09'55") 2045 B.C. (XII th Dynasty)
Ancient Chinese
Ancient Hindu
Thales
Pytheas
Eratosthenes
Hipparchus
Ptolemy
Chinese
Al Battani
Al Mamun
Arabic
Ibn Mouran
Chinese
Ibn Junis
Alpetragii
Nasseddin
Alphonsine Tables
Chinese
G. de St Cloud
Prophatius Judaeus
J. de Liguieres
Ibn al Shatir
Ulugh Beigh
Toscanelli
Walther
Navarre
Werner
Copernicus
Cassini
Manfredi
Purbach
Regiomontanus
Bianchini
Nasireddin Bic
Cassini
Spencer Jones
Dyson
Christie
Cassini
Airy
Laverrier
Bessel
Arago
Mathieu
Brinkley
Pond
Delambre
Piazzi
Maskelyne
Mayer
Bradley
Lacaille
Condamine
Godin
Bouguer
Cassini
Flamsteed
Marseilbaur
La Hire
Hevelius
Riccioli
Gassendi
Peirec
Kepler
Mendelin
Wright
Tycho Brahe
Ignatius Danfe
Treve al Hess
Rothmann
Byrge
OBLIQUITY OF ECLIPTIC
CURVES DERIVED FROM: 1. NEWCOMB'S FORMULA
2. OBSERVATIONS

Notes and References

[1] In astronomy, *obliquity* is the inclination angle of a planet's rotational axis in relation to the perpendicular to its orbital plane. It is sometimes also called *axial inclination* or *axial tilt*. The obliquity is expressed as the angle made by the planet's axis and a line drawn through the planet's centre perpendicular to the orbital plane. For the earth this is currently about 23° 27'. Long term changes in this angle are termed secular variations to distinguish them from those that occur on shorter cycles.

[2] Dodwell's unpublished manuscript "The Truth of the Bible" is in the Mortlock Library at the University of Adelaide.; Barry Setterfield, late of the Astronomical Society of South Australia, attempted to have the manuscript published in the early 1960s, without success — see: www.setterfield.org/bio.html

[3] I first heard of Dodwell's work in the mid-1980's via second hand reports — and I did not realise its possible significance for my own theories at that time. I later received a letter from a reader of my own book noting some apparent similarities. [see: *The Atlantis Researches: the Earth's Rotation in Mythology and Prehistory* Third Millennium Publishing, (1995)]

[4] For a highly entertaining summary of the Dodwell manuscript, see that given by V. Lloyd in:
 www.adam.com.au/bstett/SkepticsObliquityEcliptic33.htm

[5] Norbergen, R, *Secrets of the Lost Races*, New English Library, London (1977), pp 20-21.

[6] Dunbavin, P, *Under Ancient Skies, Ancient Astronomy and Terrestrial Catastrophism*, Third Millennium Publishing, Nottingham (2005)

[7] This is unfortunate, but understandable. It seems there was no way that I could win. To have his unique theories criticised by the astronomical establishment was clearly undesirable; but to have them praised by a 'catastrophist' would surely be even more detrimental to his reputation. As I do not use aeroplanes and unable to travel to Australia, I am therefore indebted to Moe Mandelkher for supplying to me a copy of the manuscript from a USA source; and also to Philip Clapham for directing me to the earlier SIS discussion of Dodwell's work.

[8] Various Internet references (c.2005) illustrate this usage:
      www.creationism.org/ackerman/AckermanYoungWorldChap11.htm;
      www.grazian-archive.com/quantavolution/ QuantaHTML/ vol_04/
         lately_tortured_earth_04 .htm;

www.iidb.org/vbb/archive/index.php;
tccsa.tc/archives/debate/rso_vs_lawson_9_2000.html;
gondwanaresearch.com/hp/set.htm.

[9] www.enlightened.org.uk/science.html; The Science of Creation

[10] M. M. Mandelkehr, *An Integrated Model for an Earthwide Event at 2300 BC, Part I: The Archaeological Evidence*, S.I.S. Review, Vol. V:3, pp. 77-95 (1983). Also *Part II: The Climatological Evidence*, C & C Review, Vol. IX, pp.34-44 (1987). Also *Part III: The Geological Evidence*, C & C Review, Vol. X, pp.11-22 (1987)

[11] MM Mandelkehr, *The 2300 Event : Archaeology and Geophysics. The Meteoroid Stream*, published by Outskirts Press of Denver, Colorado:2006.

[12] Clube, V. And Napier, W., *The Cosmic Serpent*, Faber & Faber, London, (1982).

[13] My own proposal concerned a wobble commencing around 3100 BC and decaying exponentially by about 500 BC.

[14] SIS Review Vol V No 2 (1980/81)

[15] SIS "Ages in Chaos" conference, 1978, p78

[16] Dodwell's conclusions were based upon the work of Sir Norman Lockyer and a 1921 survey of Karnak by F.S. Richards: Survey of Egypt Paper No 38.

[17] According to Dodwell the calculation of this line was based on:

$$\varepsilon\ 1900+T = 23°\ 27'\ 8".26 - 46".845T - 0".0059T^2 - 0".0181T^3$$

from Newcomb's formula, where T is in Julian centuries from the epoch 1900. For further information see: Newcomb, S.: 1906, *A Compendium of Spherical Astronomy*, Dover Publications., New York, 1960 (pp 226-238).

[18] Dodwell cites H. Crabtree, *Spinning Tops and Gyroscopic Motion*, 1909, pages 14 and 124.

[19] Dodwell took a mean value for these historical dates from contemporary versions of the Cambridge Ancient History, Breasted and Budge.

[20] Chronology of Ancient Egypt - Dr_ Zahi Hawass.htm

[21] Lockyer, J.N., *The Dawn of Astronomy*, Macmillan, London (1894), pp 99-120.

[22] Gerald Hawkins in 1974 resurveyed the Karnak alignment and concluded that the Amon-Re temple aligned precisely with the rising of the midwinter sun when it was rebuilt by Tuthmosis III. See. Hawkins, G.S. Phil. Trans. R. Soc. London. Series A, Mathematical and Physical Sciences, Vol. 276, No. 1257, *The Place of Astronomy in the Ancient World* (May 2, 1974), pp. 157-167.

[23] Reese, R.L., Sky and telescope (March 1992), *Midwinter Sunrise at El*

Karnak, pp 276-278

[24] Another recent study suggests that Egyptian temples were typically oriented perpendicular to the Nile and that Karnak may have been chosen because it is the only site where the winter solstice sunrise is perpendicular to the river. See Shaltout, M & Belmonte, J., JHA 2005, *On the orientation of Egyptian Temples I: upper Egypt and Lower Nubia*; www.iac.es/folleto/research/preprints/files/PP05003.pdf.

[25] Strabo, Geography, I. 4. 4

[26] On this see V. Lloyd in: www.adam.com.au/bstett/SkepticsObliquityEcliptic33.htm; he tells us: "Mr Dodwell also sent a copy of his manuscript to the Royal Society but this august body decided against publication on the grounds that 'errors of ancient observations needed further discussion'."

[27] Dodwell used the translation by C.P.S. Menon, *Early Astronomy and Cosmology, Astronomical data from the Jaina Astronomical Treatise called Suryaprajnapti* (undated).

[28] Dodwell cites the French translation of the Chou-pei by Eduard Biot in the *Journal Asiatique*, Paris (1841)

[29] Ray, T.P., *The winter solstice phenomenon at Newgrange, Ireland: accident or design*, Nature, 337, 343-5. (1989) p344. The range of azimuths for which the beam would shine down the Newgrange passage are given as between -22° 58′ and -25° 53′.

[30] ibid, p 345

[31] Toomre, A., *On the 'Nearly Diurnal Wobble' of the Earth*, Geophys, J. R. Astr. Soc., 38, 335-348 (1974)

[32] Capitain, N. & Xiao, N., *Some terms of nutation derived from the BIH data*, Geophys, J.R. astr. Soc 68, 805-814 (1982)

[33] Jeffreys Sir H, *The Earth*, Cambridge University Press (editions 1924, 1929, 1952, 1959 through to 1976).

[34] I have examined the consequences of this motion more thoroughly in chapter 10 of *The Atlantis Researches*, which was republished, with additional notes as: *Atlantis of The West*, Constable & Robinson, London (2002).

[35] Lieske, J.H. et al: 1977, "Expressions for the Precession Quantities Based Upon the IAU (1976) System of Astronomical constants.", Astronomy & Astrophysics 58,1

[36] The International Astronomical Union has now adopted an obliquity of 23° 26' 21.44" based on an epoch of 2000 as a base for such calculations.

8 Akhenaten, Eclipses and the Chronology of the Egyptian XVIII Dynasty

Summary: *According to the conventional chronology for the Egyptian VXIII Dynasty, this period coincides with a statistically rare concentration of total eclipses visible from Egypt. In addition, a boundary stela from Amarna links the building of the city to a series of 'evil omens'. This paper considers whether these omens could have been the eclipses; and discusses the consequences both for the conventional chronology of this period and the evidence for changes to the Earth's rotation.*

Many would view the Eighteenth Dynasty pharaoh Akhenaten as a great religious reformer, the first monotheist; others might prefer to see him as a tyrant. It is known that in the first year of his reign, Amenhotep IV, as he then was, abandoned the old Egyptian gods and declared that henceforth he would worship solely the Aten: an aspect of the sun-disk. In the fourth year of his reign he renamed himself *Akhenaten*, or 'Spirit of the Aten' and proclaimed Atenism as the official state religion, shunning Amun and the all the old gods. What is most surprising is that the Egyptian people went along with this heresy and indeed, it would persist into the reign of his successor Tutankhamun.

Towards the end of the reign of his father Amenhotep III, the young Amenhotep IV was installed as co-regent. A record of the ceremony has come down to us in a temple inscription at Karnak, dedicated to the god *Re-Horakhti* – Re-in-the-Horizon. The inscription describes the new deity as: '*Re Horakhti who rejoices on the horizon in his name of Solar light (Shu) which appears in the Solar Globe (Aten)*'.[1]

In the Dream Stela, positioned between the paws of the Sphinx, Akhenaten's grandfather Thutmose IV had claimed that *Horemakhet* 'Horus-in-the-Horizon' had spoken to him in a dream; he would make him king, if he would only remove the sand from around his paws! That this was the Aten is further confirmed by a commemorative scarab in the British Museum, with an inscription that he had led his army to subjugate the land of the Mitanni (Syria and north Palestine) 'with Aten before him'.[2] Here we glimpse an early form of the Atenist religion that Akhenaten would later extend.

In the fifth year of his reign, Akhenaten relocated his entire court to a new capital city at Tel-el-Amarna on the eastern bank of the Nile. He named his new city *Akhetaten*, "horizon of the Aten". The location was significant because at Amarna, the sun could be seen to rise in a cleft in the eastern mountains, which does look rather like the *Akhet* hieroglyph (left), depicting the sun rising (or setting) between two peaks. An inscription on a boundary stele, declares that the site was chosen by the Sun God himself, '*in the place which the Aten enclosed on the eastern bank for His own self*'.[3] A boundary stela dated conclusively to year 5 of Akhenaten, gives us again the various titles of the new deity: '*...Rē-Herakhte, rejoicing in the Horizon in his aspect of the Light which is in the sun-disk...*'[4] Before Akhenaten the Aten was regarded as simply the disk in which the Sun God resided, in various forms, such as Re or Shu; after Akhenaten the sun-disk absorbed all these aspects and became itself the god.

Akhenaten's heresy was abandoned after only eighteen years by his successor, the boy-king Tutankhaten, who changed his name to Tutankhamun and returned the royal court to Thebes. Horemheb would later deface Akhenaten's statues and inscriptions; and his name would be struck from later king lists. The city of Akhetaten was demolished. Of the Aten, we subsequently hear no more.

A conventional dating of Akhenaten's reign (omitting co-regencies), is as follows:

Thutmose IV	1392-1382 BC	10 years
Amenhotep III	1382-1352 BC	30 years
Amenhotep IV-Akhenaten	1352-1336 BC	18 years
Tutankhamun	1336-1327 BC	9 years

The precise dates are a subject of ongoing debate. The length of Akhenaten's reign is also uncertain, as it is not known whether he was co-regent with Smenkhkare for the last two years of his reign; or whether this little-known king reigned alone.

The present author has researched the occurrence of eclipses across Egypt during this period and it may be seen that, on all generally accepted criteria, Egypt experienced an unusually high incidence of total eclipses during this dynasty.[5]

If the diurnal rotation were constant then the location from which historical eclipses were viewed could be retro calculated with certainty. However, the tidal retardation of the rotation means that the eclipse track will fall further east than expected. Astronomers refer to this small discrepancy (TDT - UT) as ΔT (delta-T). The cumulative discrepancy between clock and calendar due to this slowing has been calculated at about half a day since 1500 BC.[6] Therefore, any eclipse track of this era would actually fall on the opposite side of the world unless appropriate adjustment is made for ΔT. There remains a range of statistical uncertainty in the value of ΔT (about 15 minutes) and to fix the time with more certainty the astronomers would need a historically dateable report of an eclipse. However, the Egyptologists would like to fix historical events precisely by a reference to the astronomical event. Inevitably this becomes something of a circular argument.

The currently accepted values for ΔT show that a total eclipse crossed the Nile on *August 15 1352 BC* (ΔT=35701.4) and would have been total at Amarna.[7] According to the conventional chronology, this was the first year of Amenhotep IV/ Akhenaten.

However, on *May 14 1338 BC* (ΔT=35352.8) another long eclipse crossed the upper Nile at Aswan and may just have been total at Thebes.[8] At just under seven minutes it came close to the maximum magnitude possible for a total eclipse. The conventional chronology puts this towards the end of Akhenaten's reign. Only four years later, on *December 30 1332 BC* (ΔT=35185.3) another eclipse was visible from Egypt.[9] However, this was a dawn eclipse. From a location midway between Amarna and Cairo, it would have risen only partially eclipsed and totality occurred just above the horizon. According to the conventional chronology, this was the fourth year of Tutankhamun's reign.

As this was a period when the sun-disk was the sole deity throughout Egypt, we should ask *why* these eclipses were not clearly recorded in inscriptions. During the Amarna period, we find numerous illustrations from the city of Akhetaten. These depict the sun-disk casting its rays down upon the favoured king and his family as they worshipped the Aten.

The apparent absence of any reference to these eclipses within contemporary inscriptions therefore raises the possibility that the conventional chronology is incorrect. This is a possibility that will no doubt appeal to the ultra-revisionists; and a question that conventional Egyptologists cannot simply ignore! For example, according to the new chronology promoted by David Rohl, the reign of Akhenaten would fall some 377 years later, between 1022 BC and 1006 BC.[10] The eclipse dates of course, remain unchanged, and on Rohl's analysis, they occurred during the Second Intermediate or 'Hyksos' period − a time from which little record of anything has survived.

Alternatively, the apparent lack of any references to the eclipses raises the possibility that the Atenist heresy could itself have been inspired by this series of prominent eclipses; and that Egyptologists just do not recognise the references.

Neither is it entirely clear what we should be looking for. Egyptologists do not know what the hieroglyph for an eclipse should look like. In a series of papers on the *Eclipse Chasers* website, the Egyptian astronomer Aymen Mohamed Ibrahem has proposed some interesting ideas.[11] He argues that the symbol *Akhet*, long thought to mean 'horizon' may sometimes stand for 'solar eclipse'; or more literally he believes '*Akhet net Pet*' to stand for 'horizon of heaven': as if the sun sets and rises on a heavenly horizon during an eclipse.[12] Another of his theories relates to the Sphinx: the Sphinx was the Egyptian lord of solar eclipses and its name '*Horemakhet*' would mean literally: 'Hor in the eclipse'. On this interpretation, the city of '*Akhetaten*' would therefore stand for 'eclipse of the sun disk'; perhaps telling us that it marks the place where a solar eclipse was viewed.

Another of Ibrahem's conclusions relate to the eclipse of June 24 1312 BC (ΔT=34695.4) which passed over Anatolia. This may have been recorded in a Hittite text and the synchronisms provided by the Amarna letters have therefore enabled him to propose a revised chronology. However, on his interpretation, it is the eclipse of 1352 BC that fell during year 4 of Akhenaten and therefore inspired the building of the city.[13]

The eclipse of 1312 BC is certainly pertinent to these investigations. It seems it occurred during the preparations for departure of the Hittite army from its winter camp. The precise site of this camp cannot be determined and neither can the date of departure, but Ibrahem convincingly argues that the 1312 BC eclipse is the only one that fits with the conventional dating – although he agrees that June might be rather late for the army to commence its campaigns.

The present author is sceptical of any major revisions to the standard Egyptian chronology. However, one possibility is that the conventional chronology could be just a few years in error; and only a small difference in the calculated value of ΔT (well within the statistical margin of error) would suffice to make the dawn eclipse of 1332 BC total at Amarna. In my own recent book *Under Ancient Skies*, I wondered if Akhenaten himself could have been there to see it. Alternatively, it might have been reports of the blackened sun rising in the gap in the mountains which led Akhenaten to believe that Amarna was the site chosen by the god himself?

It is also possible, as we have seen above, that if the dawn eclipse of 1332 BC fell in year 4 of Akhenaten's reign, then the king could already have experienced two previous total eclipses during his lifetime. Most people will be fortunate to see even one eclipse unless they chase them around the world! This common experience by the Egyptian people would at least explain why Akhenaten thought he was the chosen one; and also why the priests at first went along with his heresy.

If Egyptologists could agree upon an unambiguous inscription that records one of these eclipses then not only would it fix the Egyptian chronology, but it would also establish the stability of the Earth's rotation back to the fourteenth century BC.

A boundary stela from the city of Akhetaten holds a fragmentary proclamation by Akhenaten, which Egyptologist Cyril Aldred translated as follows:

...as Father Aten lived, something had been said which was more evil than that which the king had heard in his Year 4...more evil than what he had heard in his year 1...more evil than what King (Amenhotep III?) had heard...more evil than what king Tuthmosis IV had heard...[14]

In *Under Ancient Skies*, I left unanswered the question of what king Thutmose IV might have heard? Perhaps it too, was a report of a prominent eclipse – but I was unable to find one that would fit the requirements during the conventional dating of his reign.

According to the scarab inscription, Thutmose IV led his army east into the land of the Mitanni with the Aten before him. If this is merely the rising sun then what is so remarkable? The sun rises every day in the East. Could this be another reference to a dawn eclipse? There is one that might fit this requirement – but this would require a deviation from the standard calculation of ΔT.

A retro calculation using the accepted value for ΔT of 36295.9 seconds shows a dawn total eclipse occurring in Iran on *May 3 1375BC*, just southeast of modern Tehran.[15] However if we instead posit a ΔT of 33000 seconds then its track falls further west in Northern Syria – the land of the Mitanni. The next question is: what effect would this revised value for ΔT have upon the other eclipses discussed above?

It may be seen that if the tracks of the other eclipses are adjusted in proportion, then the track of the 1352 BC total eclipse would cross the Nile further south near Aswan. The eclipse of 1338 BC would have been seen much further north and should have been total at Thebes. The 'dawn' eclipse of 1332 BC however, now falls over the Libyan coast, and the narrow path of totality crosses the Nile further south, indeed not far from the point where the 1352 BC eclipse also crossed the river. So we may see that all three eclipses would remain visible from Egypt during the Amarna period, even on this revised scenario.

The track of the total eclipse of 1312 BC, being more east-to-west is scarcely affected, and is shifted only slightly further north, but closer to the Hittite capital of Boghazkoy. Since we do not know the precise location of the Hittite winter camp, this cannot harm the hypotheses discussed in any way. An element of statistical uncertainty remains in all of these retro calculations. There is little point in being pedantic as to the precise tracks, until such time as an inscription is found that supplies both an exact location and time of day for one of these eclipses.

So, on these astronomical arguments it is now possible to suggest a chronology for the Egyptian XVIII Dynasty. Consider:

- If the omen of year 4 of Akhenaten was the last of 4 eclipses, then it must have been that of 1332 BC;
- Therefore year 1 of Akhenaten was four years earlier in 1336 BC.
- Therefore the omen of year 1 was a report of the 1338 BC eclipse.
- The year 1352 BC falls within the long reign of Amenhotep III
- Amenhotep III reigned between approximately 1366 BC and 1336 BC

- The year 1375 BC falls shortly before the reign of Thutmose IV (as the eclipse prophesied his accession).
- Thutmose IV reigned between 1374 BC and 1366 BC

However, since we can be confident that the eclipse of 9 May 1012 BC was viewed at sunset in Ugarit on the Syrian coast; this implies that the standard value of ΔT holds for that date. This eclipse was discussed by Rohl in support of his new chronology;[16] and the present author finds little cause to disagree with this identification.[17] This would suggest that the postulated disturbance of the Earth's rotation must have occurred *sometime between* 1312 BC and 1012 BC, most likely during the 'dark age' that has long been referred-to as the Third Intermediate Period. It suggests that some event may have altered the length of day slightly.

For adherents of Velikovsky, who are accustomed to the idea of catastrophic upheavals, the possibility that a glitch in the diurnal rotation may have made the earth rotate about an hour faster over a period of 300 years, will not seem very significant. However, even this modest supposition would not be welcomed by conventional scholarship. So upon this subject I shall speculate no further beyond the evidence presented above, for it relies crucially upon the identification of an eclipse that may have been seen by Thutmose IV; and upon a correct interpretation of the extraordinary events of the Amarna period.

** 2020 Additional Note

In parallel with my own eclipse investigations: 1998-2005 for *Under Ancient Skies* the following paper by William Murray came out in 2003. It represents genuine parallel thinking on the same subject:

www.egyptologyforum.org/EMP/DAPE.pdff

Eclipse tracks with standard delta-T

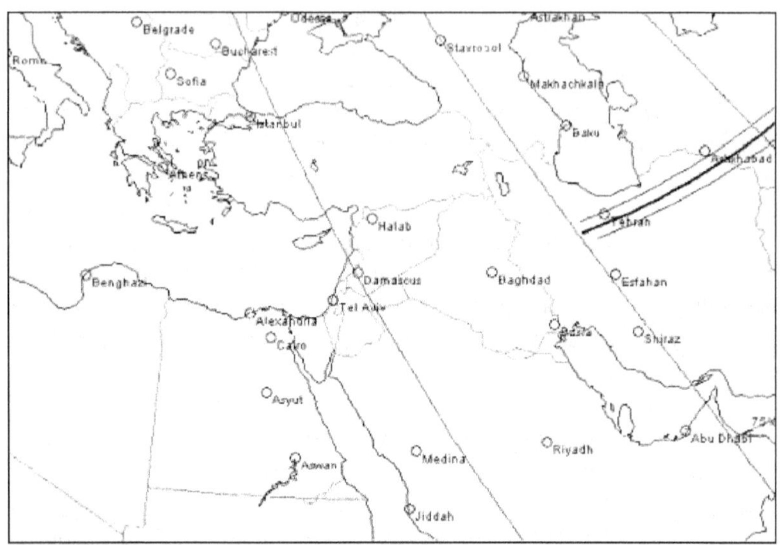

1375 BC 03 May

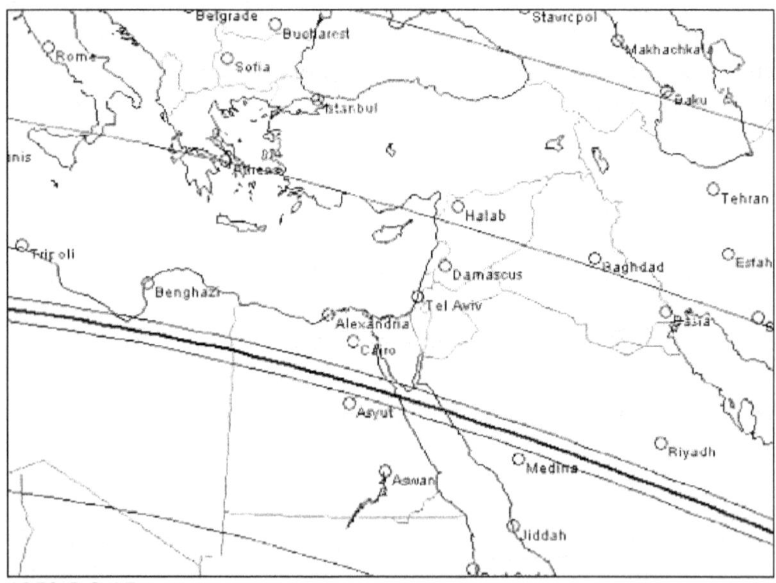

1352 BC 15 August

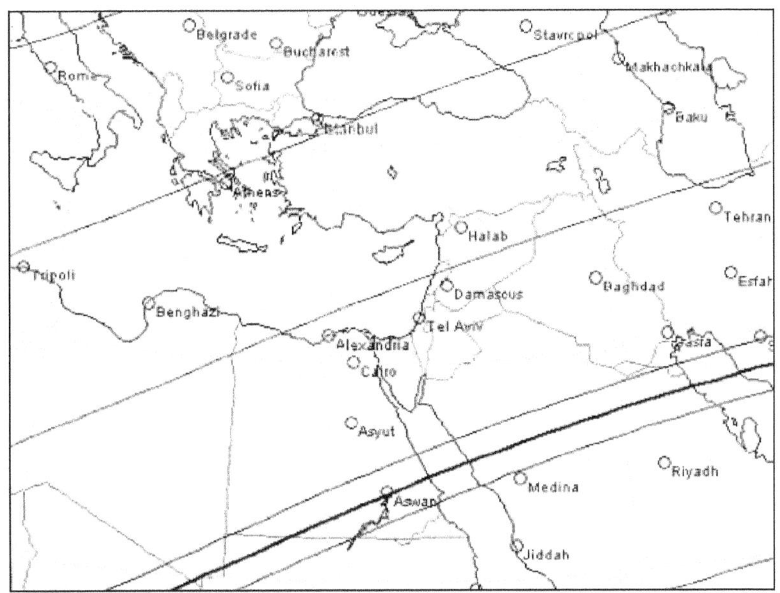

1338 BC 14 May

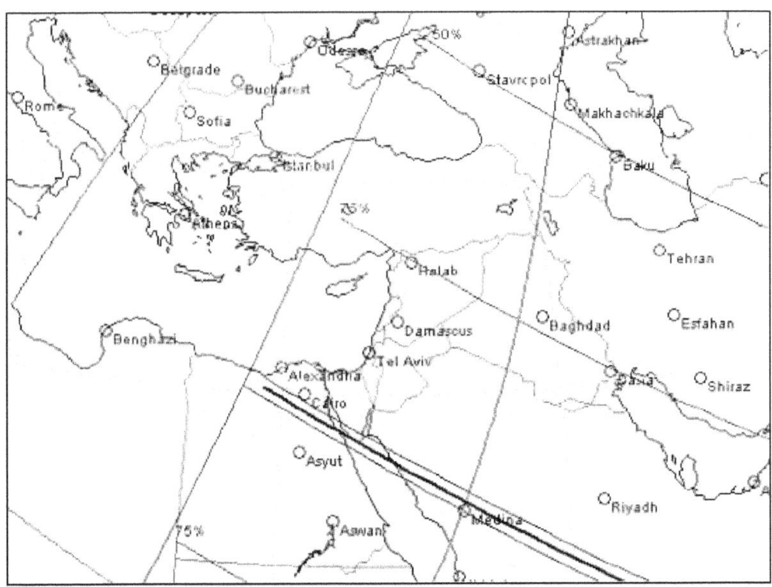

1332 BC 30 December

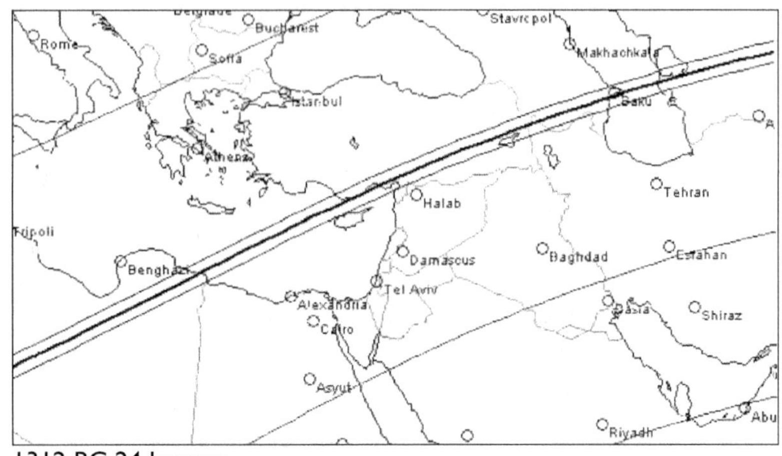

1312 BC 24 January

****

Eclipse tracks using an experimental delta-T of 33000 seconds

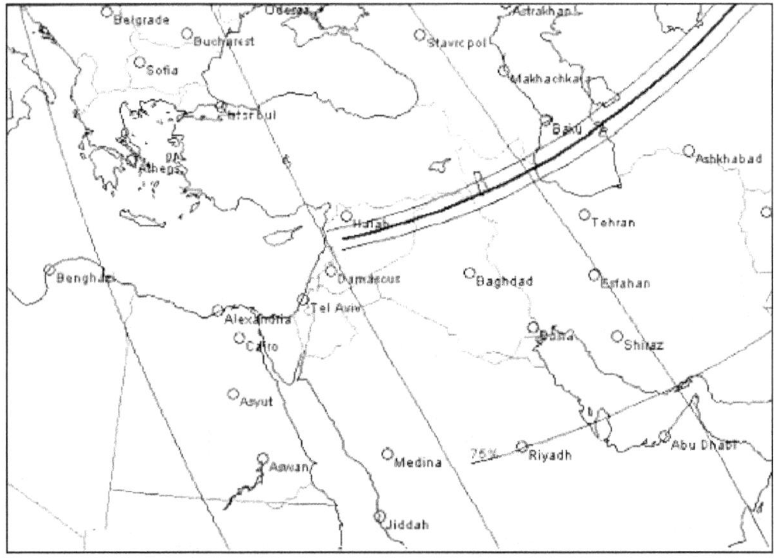

1375 BC 3 May

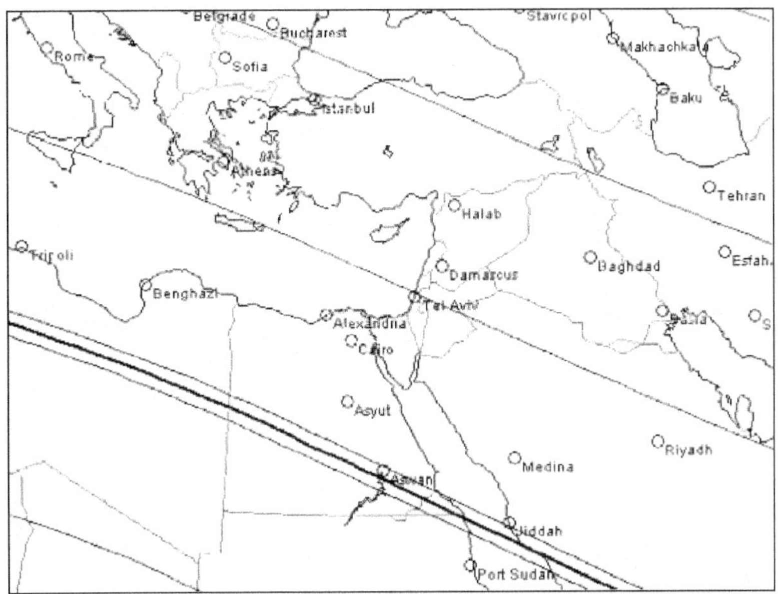

1352 BC 15 August

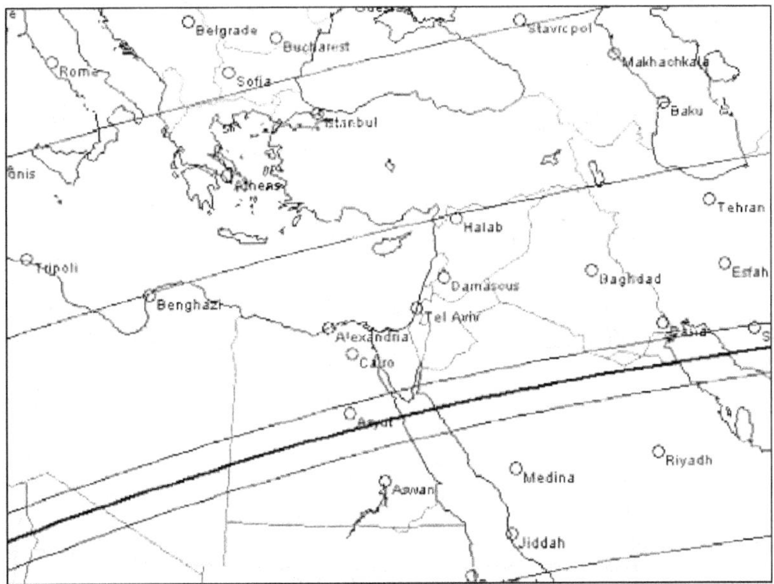

1338 BC 14 May

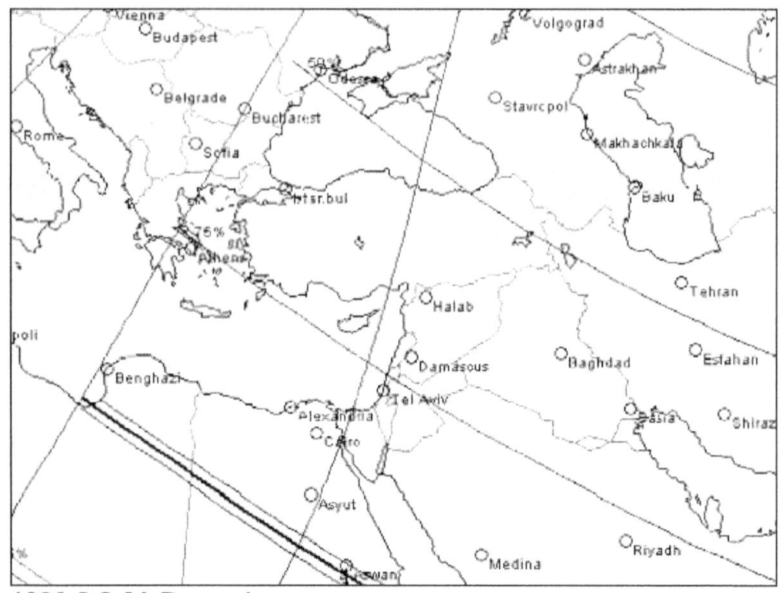

1332 BC 30 December

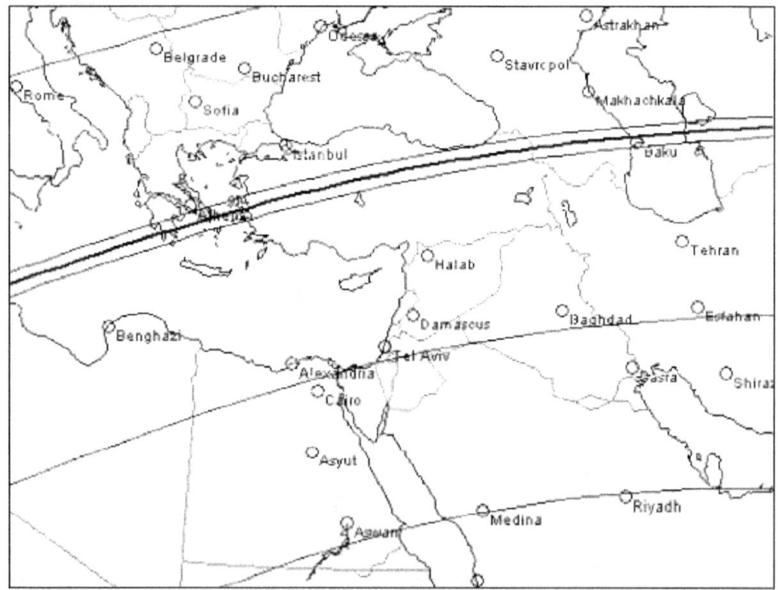

1312 BC 24 June

Notes and References

[1] Translation given by Desroches-Noblecourt, C., *Tutankhamen*, Penguin, Harmondsworth (1965), p 92

[2] Ibid, p 26

[3] Aldred, C., *Akhenaten, King of Egypt*, London (1988), p 49

[4] Ibid, p 47

[5] On average, any location on the earth's surface should experience an eclipse about every 375 years. See: Meeus, J., J. Brit. Astr. Assn, 92, 124-6 (1982)

[6] Stephenson F.R. and Holden M.A. *Atlas of Historical Eclipse maps*, Cambridge University press (1986) introduction, iix

[7] Retro-calculations based on Skymap Pro 6

[8] Retro-calculations based on Skymap Pro 6

[9] Retro-calculations based on Skymap Pro 6

[10] Rohl, D., *A Test of Time*, Century, London (1995), P 241

[11] http://www.eclipse-chasers.com/egygod1.htm

[12] Aymen Ibrahem, *Egyptian Cosmology, Part VII, Karnak the Horizon of Heaven*, (2000): http://members.aol.com/KCStarguy/blacksun/egyptianeclipse.htm

[13] http://eclipse-chasers.com/egypt4.htm

[14] Aldred, C., *Akhenaten, King of Egypt*, London (1988), p 50

[15] Retro-calculations based on Skymap Pro 6

[16] Rohl, D., *A Test of Time*, Century, London (1995),

[17] Dunbavin, P. *Under Ancient Skies: Ancient Astronomy and Terrestrial Catastrophism*, Third Millennium Publishing, Nottingham (2005), p 133.

9   Raised Beaches & Submerged Forests –
Curious Anomalies

Summary: *The following analysis of Holocene sea-level changes is intended as an overview for the general reader. Use of specialist terminology has therefore been held to a minimum and is defined where its use is unavoidable. The purpose is to question some of the assumptions that underlie modern theories about the causes of sea-level change.*

Under gradualist geology there is no mechanism that properly explains the formation of raised beach terraces. For the sea to cut cliffs and form beaches at the shoreline requires a stable sea level to prevail for hundreds of years and then to *rapidly fall* to cut a terrace at a lower level, without leaving transitional features. If sea level were gradually changing over time, as the glaciologists tell us, then no intermediate beach terraces would be able to form.

Also around modern coastlines are sometimes found the 'submerged forests'; a term that applies to any form of land-based vegetation found below beach sand. Usually however, it is the presence of mature tree roots, still in their position of growth that draws attention to such features. Young trees cannot grow in salt water, so again it implies a *rapid rise* of the sea to submerge the mature woodland and bury the roots with sand. Gradual flooding would have allowed them to be eroded away by the tides.

The standard theory of climate and sea-level change, as it has evolved, need only be summarised. It represents primarily 'western' science, from surveys by American, European and Australian fieldworkers which are then extrapolated to give a world view. On this theory, over the past million years, there have been some twelve glaciations; the most recent cold period, often simply referred-to as 'The Ice Age', reached its peak around 18,000 years ago before rapid warming to the interglacial *Holocene* epoch around 11,700 years ago. The text-book consensus is that at the peak of this most recent glaciation, the North American 'Laurentide' ice-sheet expanded to a depth of some 3,500 metres as it spread across Canada and Greenland, and the northern ice-cap further extended across northern Britain, Scandinavia and on into the Arctic Ocean.

In the 1960s when the Milankovich astronomical theory of ice ages gained general acceptance then it took control of all theories of past climate and sea-level. It offered a process acceptable to the gradualist consensus. For regions such as the Canadian arctic, Scandinavia and Northern Britain, where the scouring of former ice-sheets and the apparent rebound of the land is visible in the landscape, this leads to the simple equivalence of where did all this melt-water go? Plainly it went into raising the sea-level worldwide, perhaps by as much as 120 metres; often we find this referred-to as the 'post-glacial sea-level rise' and its drowning of the coasts as the 'Holocene Transgression'. At the last glacial maximum (LGM) around 18000 BP sea levels were at their lowest and since then have risen as the ice-caps melted. These broad outlines and their astronomical causes are not challenged here; the question is whether there were also other forces at work.

Among many consequences, the post-glacial sea-level rise explains the drowning of the North-European continental shelf, separating Britain and Ireland from the rest of the continent. The shallow North Sea basin must have been dry land when so much of the world's water was locked up as ice. This was further confirmed when trawlers began to drag-up animal bones and flint arrowheads. Conventional dating for the submergence of this former tundra region follows the glacio-eustatic theory, suggesting that it was drowned around 8500 BP when the land-bridge between Britain and the continent was finally severed. The shallowest part of the North Sea: the Dogger Bank, may have lingered on until the soft sediments were eroded away like so much of the modern east coast of England. The constraints of glacio-eustatic theory therefore dictate that similar low-lying land around Britain and Ireland must similarly have been drowned around the same time. The drowned forests found below Welsh and Irish beaches suggest that the submergence stabilized around 5,000 years ago. A prime example is that at Borth, mid-Wales.[1] For a pictorial view of this ancient forest see:

https://www.mirror.co.uk/news/uk-news/storm-hannah-unearths-sunken-forest-16201397

The submerged forest at Borth was dated to c.5300-5100 BP with other examples similarly but less securely dated around the British coasts.[2] Raised beach terraces of Holocene age are not found around North Atlantic coasts.

Across arctic Canada and Scandinavia the creation of a sequence of raised-beach terraces can be convincingly explained by the rebound of the land since the removal of the ice burden; a process termed 'glacial isostasy' together with 'glacial eustasy' for its effect on the sea level. The specialists assert that the Milankovich orbital factors, together with glacial eustasy and isostasy, can explain all of the variations. There is also a tendency, in any field of academic excellence, for researchers to go out into the field and discover what they are expected to find. So much of modern theory is based on theoretical 'modelling'. To avoid text book bias, it is necessary to strip-out such preconceptions. Of value is solely the 'pure' dating evidence from samples collected by fieldworkers and their height relative to sea-level. It is often difficult to isolate the pure unbiased data in the specialist studies. As an example of the difficulties, see the paper by Kaplin-et-al (1993) which sought to reconcile various sea-level data across the vast area of Russia and Siberia. [3]

However, the key assumption that a vast ice sheet existed in the High Arctic of Siberia and Beringia throughout the LGM and then melted-away has been called into question by recent research in the Chukchi Sea.[4] The primary focus was to investigate transgressions dating from earlier epochs, the most recent being the Krasny-Flagian Transgression (named after the "red-flag" valley on Wrangel Island where it is extant) and dated 64,000-73,000 years ago. The reasonable conclusion is that no more-recent ice sheet could have existed in this area or it should have destroyed these older features. The Chukchi Shelf was ice-free throughout the LGM and also into the Holocene, as is evident from the survival of dwarf mammoths on Wrangel until c.2500 BC.[5] Such creatures must have been isolated there for thousands of years for the dwarf species to evolve.

More recent studies would put the eastward limit of any glacio-isostatic rebound at the longitude of Novaya Zemlya and Franz Josef Land and therefore the supposed northern ice-sheet was confined to the Barents Sea and westwards.[6] The northern ice-cap was decidedly lop-sided and centred upon northern Greenland. How Siberia – today the northern-hemisphere's coldest region – could remain ice-free, is an enigma. As a digression, it has been suggested that Kerguelen Island, at the antipodes of Greenland, may have been completely covered by an ice sheet throughout the LGM;[7] another enigma that is unexplained by conventional theories.

Other indicators also come from the far-east. The conclusion of Kaplin-et-al was that the 'Holocene transgression' created beach terraces above present day in Chukotka (far-east Siberia) and Primorye (Amur coast) but apparently not in Kamchatka, suggesting "different tectonic regimes".[8] The data itself is highly confusing. For example, a beach terrace 5-7 m above sea level in Krest Bay, Chukotka was only loosely dated "from 8-5 ka BP";[9] these are contemporary with the so-called Jomon terraces of Japan.

Shell middens of the indigenous Jomon culture may be found on raised beach sites around modern Japanese coasts. The Jomon were hunter-gatherers and their middens – heaps of discarded shells – indicate where they lived. This is quite perverse compared to the situation in Europe and America, where the coasts of equivalent age have been drowned beneath the 'post-glacial sea-level rise' and we find instead the submerged forests; but for Europe one is only permitted to whisper about the possibility of Neolithic and Mesolithic human settlements submerged offshore.

Japanese archaeologists classify the Jomon culture into various eras. The first people arrived around 12,000 years ago when the islands were joined to the Asian mainland; they may have been related to the Ainu of modern Hokkaido. The Initial Jomon is dated 8000–5000 BC. By this period, the rising sea had separated the southern islands of Shikoku and Kyushu from the main island of Honshu. The Early Jomon (ca. 5000–2500 BC) is defined by shell-middens of a coastal fishing culture and pottery similar to neighbouring Korea. The later Jomon phases up to 300 BC show a transition to Neolithic, before rice-farmers from mainland Asia displaced the native people.

At the LGM around 18000 BP, when world sea-levels were at their lowest, it follows that all the islands of Japan were linked together. Thereafter the sea-level rose in parallel with other parts of the world, reaching a maximum around 6000 BP (c.4000 BC) after which it apparently reversed. Specialists term this the 'Jomon Transgression' and the apparent decline of sea-level as the 'Jomon Regression'. To explain this regression is a puzzle for geomorphologists; they have to resort to local tectonic forces raising the land; or redistribution of loads on the sea-bed and mantle. Once such concepts find their way into published scientific papers it is assumed that some earlier researcher has 'proven' the theory and others are expected to cite it, whereas in fact, nothing is proven.

Precise dating of raised beaches is as difficult as explaining their cause. One can only say that the re-emergence must be older than the most recent datable artefacts found on them and younger than the marine organisms found within them. The Jomon shell-middens also give an indication of the proximity of the sea, based on whether they are marine or fresh-water species. A dateable example of this is the Futatsumori Shell Midden.[10] This settlement features an extensive midden from the Early through to the Middle Jomon period (approx. 3500 – 2000 BC). The plateau lies at an altitude around 30m and covers an area of about 35 ha on the west bank of Lake Ogawara near the Pacific coast.

Another example lies at the entrance of Ogushi Bay in the Goto Islands off Kyushu. The researcher Esaka in 1967 discovered Early Jomon pottery and used this to estimate the age of the beach terrace.[11] More recent researchers used 14C dating of the shells cemented in the beach-rock to date the terrace to 5650±150 BP.[12] The researchers conclude that the sea has been comparatively stable and the terrace has been continuously above the sea during the past five thousand years.

However, such anomalous high sea-levels are not restricted to Japan. For Korea, Song-et-al conclude that the Yellow Sea stood 1-2m higher than present between 7800 and 5000 BP before falling to modern levels; but again their analysis is heavily influenced by glacio-eustatic modelling.[13]

Equivalent sequences of Holocene raised beaches are extant further south. In China, we find that the entire plain of the Huang-Ho River was first occupied by farmers only since about 5000 BP. Wang and Zhao cite "hundreds" of 14C dates from along the Chinese coast, from the Bohai Sea down to Hainan, to show that the maximum of Holocene transgression occurred between 6000-5000 BP followed by a subsequent regression, as in Japan.[14] At the mouth of the Yangze River, He-et-al identify two regressions and interruption of the coastal Hemdu culture between 6700-6300 BP and again 5600-5000 BP.[15] They conclude that the East China coastal plain was a shallow sea earlier in the Holocene. Rollet-et-al also find that a shallow estuary filled the Fuzhou basin (opposite Taiwan) during the mid-Holocene highstand;[16] and small islands existed that were occupied from 5500 BP onwards. It is from this era that seafaring to Taiwan and beyond first began (proto-Polynesians).

Two of these 'islands' of settlement in the Fuzhou Basin between 5500 and 5000 BP were situated 80 km inland of the modern Chinese coast, after which the regression of the sea allowed rapid human settlement and rice farming on the newly emerged coastal plain.[17] This is the same era at which we find the submerged forests around British and Irish coasts. If the sea can regress by this distance (by whatever process) then it could also transgress the coast by a similar amount in other places; but no specialist researcher would dare suggest that we should look 80 km offshore around Atlantic coasts to locate submerged Neolithic settlements! Such is the limitation in thinking that constrains academic research. To find such free discussions then you have to consult older coastal studies that predate the glacio-eustatic theories of the 1960s.

South of the equator, Australia is considered to be a relatively stable sea-level zone, free of tectonic and glacial influences. Studies of the Great Barrier Reef in the 1970s established that sea level had reached close to present shorelines around 6000 bp (uncorrected), confirmed by more recent discovery of older drowned reefs.[18] The extensive review by Lewis-et-al of all the dating evidence confirms this but with regional differences.[19] Australian researchers will dispute whether there was a mid-Holocene highstand as in China (it seems to depend on whether they believe tubeworms are a good indicator of sea-level) while others think there were 'oscillations'.

Further south for Tasmania there seems to be general agreement that there was no mid-Holocene highstand and that the separation from the mainland occurred quite early in the Holocene, well before the level reached current shores;[20] yet evidence of Aborigine settlement suggests they were still able to reach it by canoe until c.6000 BC, which is suggestive of lowered shorelines.[21] It should be noted that Tasmania is at latitude (c.40°S) comparable with Spain and Japan; there is no continental land at higher southern latitudes to compare with northern hemisphere coasts, or with South America.

South America lies almost exactly at the antipodes of the East Asia coast and exhibits a similar pattern of raised beach terraces along an emergent coastal plain. Coastlines along the Pacific are difficult to assess due to the severe tectonic forces uplifting the Andes, but on the Patagonian side there have been a number of recent studies and attempts to explain the raised beaches.

Along the Patagonian coast, seven terraces of various ages are identified at heights between 8m and 186m, with naturalists since the nineteenth century noting the presence of marine fossils and shells on the coastal plain. At all the locations studied the highest Holocene terrace is considered the oldest, with the lower terraces being younger regressive features. For example, the sites at Camarones and Bahia Bustamente (central Patagonia) gave dates between 5380±70 BP and 6708±40 BP for terraces 6-7m above modern sea level.[22]

A similar pattern of terraces and raised lagoons continues north as far as the Brazilian coastal plain around Rio de Janeiro, where transgression peaked between 4590 and 5100 BP.[23] However here, the terraces are evident at lower levels than in Argentina, around +2.5m. It is generally recognised that the South American beach terraces are tilted, being at their highest in the south of Patagonia.

An interesting study of the Rio de la Plata estuary is that of Prieto-et-al.[24] The estuary has taken its present form only since the regression of the sea; in earlier times the estuary extended further inland along the Paraná River (Argentina) into which the Uruguay River then flowed. The researchers consider that the River Paraná then met the sea near the modern city of Rosario, which is some 250 km upstream of the present-day tidal estuary; they cite the dating to 6296 cal BP, of a baleen whale carcass found near Baradero. The modern Paraná Delta was certainly in place by 2243 BP when there is evidence of human occupation. The researchers suggest a "rapid fall" of the sea from a maximum at 5800-5200 cal BP. Perhaps a more precise dating indicator from the same study may be found at Aroyo Solis Grande (east of Montevideo) where they suggest that salt-marsh vegetation shows the highstand at 6300-5100 BP followed by a transition to brackish freshwater marsh between 5100 BP and 2900 BP.

Comparisons with China and Japan are significant here; we see, on opposite sides of the world at the same era, a comparable regression of the sea by many kilometres in shallow estuaries. In both regions we see the creation of new coastal plains. This correspondence must be contrasted with the formation of the submerged forests and peats around European coasts.

Take a step-back for a moment and consider what this means. There can only be one worldwide (eustatic) sea-level at any given epoch. The evidence is therefore demanding a geological process that can raise the vast area of land from Siberia down to China out of the sea by several metres and the same again for all of South America; and all this within only a matter of decades; while simultaneously land in the North Atlantic and perhaps also around Australia is sinking. In most of the sea-level studies cited (and others) one will see little reference to the fact that our planet is a dynamic rotating system; with the sea-level at any location determined by a balance of gravity versus the centrifugal rotation tending to throw it outward. The *geoid* or ellipsoid of the Earth varies by 13 miles (22 km) between the polar and equatorial radius. A variation of only minutes of arc in the polar locations would be all that is required to cause sea-level variations of the order that are observed.

There are other contemporary indicators away from the coasts that could be included. These mid-Holocene coastal events are coincident with climate changes. Pollen evidence defines the Atlantic-Sub-Boreal (warm to cooler) transition in Europe around 5,000 years ago. A fluctuation of glaciers in the Alps and the Andes also occurred at this time; the desertification of the Sahara region also began at this era. One could continue this list of mid-Holocene anomalies. The problem goes much wider than just fluctuating sea-levels.

In recent decades some new terminology has come to the fore within sea-level studies. The sites of former ice-caps are termed 'near-field', whereas those regions further away, where secondary uplift effects are perceived, are termed 'intermediate field' and 'far-field'. The proposal is that the rises and falls of the coastlines in the mid-Holocene, due to the perceived eustatic and isostatic factors, could be explained by a 'forebulge' in the mantle below the sea-bed further away from the seats of glaciation.[25] In summary, this bulge first raised the sea-bed but as the ice melted so the mantle returned to "former load centres"; this forebulge-collapse mechanism is offered to explain submergence followed by emergence; with the weight of ocean water fighting against the rebound with different results according to local conditions. While this may have some merit for South America, it is difficult to see how such a forebulge-collapse in East Asia could be explained by rebound from a non-existent ice sheet in Siberia; or indeed why raised-beach terraces of similar age are not found around Europe and America, where instead we find submergence. It cannot explain *why* the northern ice-cap was apparently centred on Greenland rather than the North Pole; still less could this mechanism explain the rapid climate swings ongoing at the same era. It seems that attempts to explain how coastlines could be simultaneously high in one region while being low in another are becoming ever-more elaborate and shrouded in jargon.

The weakness in all of the various glacio-eustatic 'models' is that estimates of loading on the sea-bed are based upon assumptions about how much ice has melted since the Pleistocene ice age; and the thickness of the ice-sheets are themselves calculated based on estimates of how much the sea-level has risen since they melted. It may be seen that the whole process goes around in a circle of assumptions.

A wide-ranging critique of the confusion and anomalies at the heart of modern sea-level research may be found at:

https://notrickszone.com/2m-higher-holocene-sea-levels/.

The website's authors argue, based on 80 published sources, that rising sea-levels today are not the result of increased carbon dioxide in the atmosphere (no comment). However, they do have a point that former warm-climate and high sea level regimes cannot be cited to make a case for it while the specialists remain so divided over the interpretation of ancient coastlines.

If one may attempt to summarise by focusing only on the 'pure' dating evidence of raised beaches and submerged forests (and set-aside the text-book models that may bias how the specialists present their results) then we may discern a pattern of sea-level change in alternate quarter-spheres *since c.6000 BP*. The precise date of such change is only as good as the samples chosen for dating by the various field-workers and published in their papers. To pin down a *precise* date for the transition would require specialist researchers to seek unambiguous dating evidence together with a more open-minded attitude from academic referees, to allow such work to be published for others to cite. Unfortunately, this is not how the academic process works. Usually, as we saw with the advent of radiocarbon and tree-ring dating, the eminent authorities will defend an established theory until some unexpected discovery renders it completely untenable.

Next Page:
The pattern of sea-level change since c.6000 BP plotted on an equal-area projection, as published by the author in 1995; each 'plus' or 'minus' sign marks a site with a published or an accepted date as to how the sea level (or the land height) has changed since that era. Inland sites where the snow or tree-line has changed (itself simply a height above sea-level) can be just as useful in establishing such a pattern. Such a quarter-sphere pattern would be indicative of *a small pole-shift* since this date, in addition to the glacio-eustatic influences. Any evidence of earlier pole-shifts during the ice-ages would now be drowned beneath the post-glacial sea-level rise. Further discussion and cross-disciplinary correspondences with sea-level evidence may be found in my books Atlantis of the West and Towers of Atlantis.

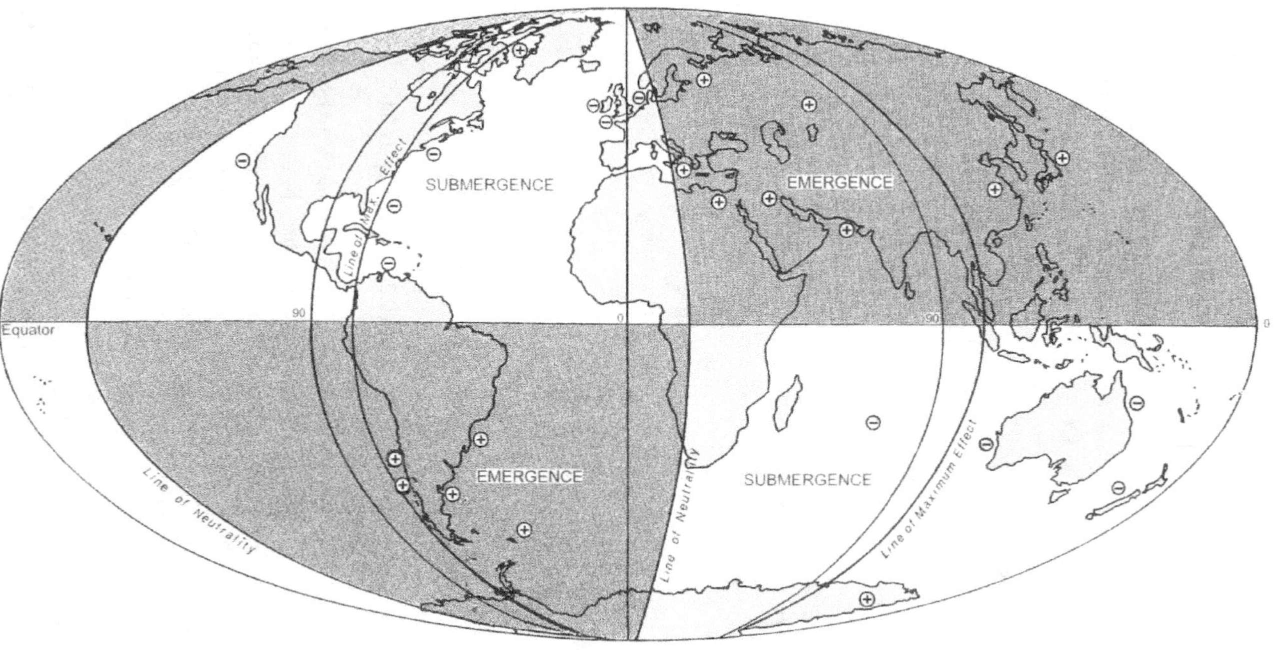

The four quarter-spheres of sea-level change around 3100 BC. This map was originally included in *The Atlantis Researches (1995)*

References:.

---

[1] Campbell, J.A. & Baxter, M.S. Radiocarbon measurements on submerged forest floating chronologies, *Nature*, 278, 409-413 (1979)

[2] Heyworth, A. Submerged forests around the British Isles, in Fletcher, J.M. (ed) *Dendrochronology in Europe*, 279-288 (1978)

[3] Kaplin P. A. Svitoch A. A. and Parunin, 0. B. Radiocarbon Chronology of Paleogeographic Events of the Late Pleistocene and Holocene in Russia in *Radiocarbon*, Vol. 35, No. 3, (1993) P. 399-407

[4] Lyn Gualtieri, Sergey Vartanyan, Julie Brigham-Grette, and Patricia M. Anderson; Pleistocene raised marine deposits on Wrangel Island, northeast Siberia and implications for the presence of an East Siberian ice sheet, in *Quaternary Research* 59 (2003) 399–410

[5] Vartanyan, S.L., Garutt, V.E., Sher, A.V., 1993. Holocene dwarf mammoths from Wrangel Island in the Siberian Arctic. *Nature* 362, 337–340.

[6] Forman, S.L et al (2010) Holocene relative sea-level history of Franz Josef Land, Russia, *Geological Society of America Bulletin* 1997;109;1116-1133.

[7] Hall, K.J.(1984) Evidence in favour of total ice cover on sub-Antarctic Kerguelen Island during the last glacial, *Paleogeogr. Paleoclimatol. Paleoecol.* 47, 225-232

[8] Kaplin-et-al, See ref 3, p.407

[9] Kaplin-et-al, see ref 3, p.403

[10]https://jomon-japan.jp/wp-content/uploads/2018/04/leaflet_14futatsumori_2018.pdf

[11]Tachibana, K. and Sakaguchi, K. Age of Beachrock containing the Jomon Pottery in the Goto Islands, The Quaternary Research (Daiyonki-Kenkyu), 1971 Volume 10 Issue 2 Pages 54-59.

[12] Ibid, citing Gak p2861

[13] Song, Bing & Yi, Sangheon & Yu, Shi-Yong & Nahm, Wook-Hyun & Lee, Jin-Young & Lim, Jaesoo & Kim, Jin-Cheul & Yang, Zhongyong & Han, Min & Jo, Kyoung-Nam & Saito, Yoshiki. (2018). Holocene relative sea-level changes inferred from multiple proxies on the west coast of South Korea. Palaeogeography, Palaeoclimatology, Palaeoecology. 10.1016/j.palaeo.2018.01.044.

[14] Wang,S and Zhao,X. (1992) Sea-level changes in China-past and future: their impact and countermeasures, In McCall G J.H. et al. (eds.), *Geohazards* pp.161-169

[15] Keyang he, -et-al, Middle-Holocene sea-level fluctuations

interrupted the developing Hemudu culture in the lower Yangtze River, China, *Quaternary Science Reviews,* May 2018

[16] Rolett, BV, Zheng, Z, Yue,Y (2011) Holocene sea-level change and the emergence of Neolithic seafaring in the Fuzhou Basin (Fujian, China), *Quaternary Science Reviews* 30 (7-8), 788-797

[17] ibid

[18] Chappell, J., Rhodes, E.G., Thom, B.G.and Wallensky, E., (1982) Hydro-Isostasy and sea-level isobase of 5500 B.P. in North Queensland, Australia. *Mar.Geol.*, 49, 81–90.

[19] Lewis, S.E. et al (2008) Mid-late Holocene sea-level variability in eastern Australia, *Terra Nova*, 20, 74–81

[20] Nakada, M. & Lambeck, K. Late Pleistocene and Holocene sea-level change in the Australian
region and mantle rheology, *Geophysical Journal* (1989) 96, 497-577

[21] Fletcher, M.S & Thomas, I. (2010) A Holocene record of sea level, vegetation, people and fire from western Tasmania, Australia, *The Holocene* 20(3) 351–361

[22] Rostami, K. Peltier, W.R. " & Mangini, A. Quaternary marine terraces, sea-level changes and uplift history of Patagonia, Argentina: comparisons with predictions of the ICE-4G (VM2) model of the global process of glacial isostatic adjustment, *Quaternary Science Reviews* 19 (2000) 1495-1525

[23] Wagner, J.A. et al, Sea-level fluctuations and coastal evolution in the state of Rio de Janeiro, southeastern Brazil, *Anais da Academia Brasileira de Ciências* (2014) 86(2) 671-683.

[24] Prieto A.R et al. Relative sea-level changes during the Holocene in the Río de la Plata, Argentina and Uruguay: A review, *Quaternary International* 442 (2017) 35-49
http://dx.doi.org/10.1016/j.quaint.2016.02.044

[25] Khan N.S. et al. Holocene Relative Sea-Level Changes from Near-, Intermediate-,and Far-Field Locations, *Curr Clim Change Rep* (2015) 1:247 – 262.

This article was originally published as an interactive web-page in 2019

https://www.third-millennium.co.uk/raised-beaches-submerged-forests

10   Submerged Forests around British and Irish Coasts

Summary:
The following article was originally published as an interactive web-page. The map below shows the principal locations where submerged forests are found around British and Irish coasts; giving hyperlinks to both popular and specialist reports; some of these will give further detail and maps together with photographs.

Radiocarbon dates are shown here in square brackets. The dates suggested in the various press reports should be treated only as an approximate guide to the precise era of submergence and so are enclosed in quotes. Trees may survive for hundreds of years and it is not always stated whether radiocarbon dated material has been taken from tree stumps in-situ or from fallen logs. Only dated wood from the final growth-rings of a stump would be indicative of the true date of submergence.

Older published studies may refer to a *'submerged forest period'* during the mid-Holocene, whereas more recent papers speak of the submerged forests as dating to various eras. This is in part, an artefact of random sampling of available wood; however it may be seen that samples from the North Sea coast tend to favour the conventional dating for the submergence of 'Doggerland', while those around western coats tend to be from a later period.

The list is not intended to be exhaustive of either sites or of links and will need to be updated as and when new sites or relevant reports emerge. The intention here is merely to give an entry point for the general reader to the photographic evidence and relevant studies.

Amble, Northumberland [7000 BP]
https://www.dailymail.co.uk/sciencetech/article-3593219/North-sea-reveals-7-000-year-old-human-footprints-ancient-forest-Woodland-stretched-Denmark-covered-ocean.html

Allonby, Cumbria [?]
See Steers, p77, also see Silloth below

Anglesey [?]
Numerous sites around the island, see
http://www.dyfedarchaeology.org.uk/lostlandscapes/submergedforests.htm
This includes a map of all the known sites around Wales

Benbecula, Western Isles ['7000 BP']
https://www.bbc.co.uk/news/uk-scotland-highlands-islands-46890793

Ballinskelligs ['4000 BP']
https://www.theringofkerry.com/blog/155-pre-historic-4-000-year-old-forest-in-ireland

Bray, Wicklow [6180 BP]
https://jasonbolton.wordpress.com/2015/04/28/the-submerged-forest-of-bray-co-wicklow/

Borth [5500 BP] and **Ynyslas** [3500 BP]
https://www.mirror.co.uk/news/uk-news/storm-hannah-unearths-sunken-forest-16201397
https://www.coflein.gov.uk/en/site/421874/details/tanybwlch-submerged-forest
https://www.nature.com/articles/278409a0

Canning Town, London [3940—3700 cal bc]
www.lamas.org.uk/images/documents/.../Pg_001-016_Submerged_forest.pdf

Cleethorpes ['Neolithic']
https://citizan.org.uk/blog/2016/Aug/25/three-submerged-forests/

Clare, Galway, Mayo [7400-5200 BP]
https://www.irishtimes.com/news/ireland/irish-news/storms-reveal-7-500-year-old-drowned-forest-on-north-galway-coastline-1.1715303

Conwy ['6000 BP']
https://www.geograph.org.uk/photo/5119512

Drigg, Cumbria [?]
See Steers p82-3

Lancashire Coast ['Neolithic']
https://citizan.org.uk/blog/2016/Aug/25/three-submerged-forests/
https://www.arnsidesilverdaleaonb.org.uk/uploads/2016/05/lsca_chapter3a.pdf

Mablethorpe, Lincolnshire ['6000 BP']
https://www.lincolnshirelive.co.uk/news/local-news/crowds-descend-submerged-ancient-woodland-1897235

Morecambe Bay ['7600-5200 BP']
see Isca report 3.11 under Lancashire

Mounts Bay, Cornwall ['4000-6000 BP']
https://www.belfasttelegraph.co.uk/news/uk/ancient-submerged-forests-uncovered-by-storm-damage-to-coastlines-30030179.html
See also the paper by French C.N. below

Norfolk ['10,000 BP']
discovered offshore by divers
https://www.bbc.co.uk/news/av/uk-england-30905267/ancient-underwater-forest-discovered-off-norfolk-coast

Orkney [4410–4325 cal BC]
https://www.scotsman.com/lifestyle-2-15039/archaeologists-survey-scotland-s-forests-under-the-sea-1-4672611
See also the detailed paper of Timpany et al, reference below.

Pentewen, Cornwall [?]
See Steers, p 258

Pett, Sussex ['Mesolithic']
https://citizan.org.uk/blog/2016/Aug/25/three-submerged-forests/

Rhyll [?]
This article includes a map of all the known sites around the Welsh coastline
http://www.dyfedarchaeology.org.uk/lostlandscapes/submergedfores
ts.html

Siloth ['8000 BP']
http://www.solwayshorestories.co.uk/shore-stories/the-submerged-
forest/

Solent [6300 BP]
Ancient logs discovered near a (recent) shipwreck site
https://www.southampton.ac.uk/~imw/ship.htm

Stolford [5398-5020 BP]
Campbell & Baxter (see Borth & Ynyslas above)

St Mary's (Scilly) [5310-5050 cal BC]
Source: *The Lyonesse Project*

Westward Ho! Devon [6500 BP]
http://www.westwardhohistory.co.uk/submerged-forest/
See also the pdf of Bell, Manning and Nayling, suggesting an earlier and a later period of submergence:
https://dendro.cornell.edu/articles/Bell2009.pdf

****

References

Campbell, J.A.; Baxter M.S. (1979). "Radiocarbon measurements on submerged forest floating chronologies". Nature. 278 (5703): 409–413. doi:10.1038/278409a0.

French, C. N., The 'Submerged Forest' palaeosols of Cornwall, Geoscience in south-west England, 9, 365-369.
www.ussher.org.uk/journal/90s/1999/documents/French_1999.pdf

Reid, C, (1913) Submerged Forests. The Cambridge Manuals of Science and Literature, Cambridge University Press. The full text is available here:
https://books.google.co.uk/books?id=OQ89AAAAIAAJ&pg=PA38&lpg=PA38&dq=Submerged+forests+Scotland&source=bl&ots=MAiHTxWLKP&sig=ACfU3U3RhYfX3g6IG4PWMNA6JNwEyRPEFA&hl=en&sa=X&ved=2ahUKEwij1Ofgg8jjAhUSWsAKHWa4A6U4ChDoATAQegQICRAB#v=onepage&q=Submerged%20forests%20Scotland&f=false

Steers, J.A. (1964) *The Coastline of England and Wales*, Cambridge Univ. Press.

Timpany, S., Crone, A., Hamilton, D., and Sharpe, M. (2017) Revealed by Waves: A Stratigraphic, Palaeoecological, and Dendrochronological investigation of a Prehistoric Oak Timber and Intertidal Peats, Bay of Ireland, West Mainland, Orkney, *The Journal of Island and Coastal Archaeology*, 0:1–25 http://dx.doi.org/10.1080/15564894.2017.1284960

Bell, M., Manning, S.W. & Nayling, N. 2009. Dating the Coastal Mesolithic of Western Britain: A Test of some Evolutionary Assumptions. In P. Combé, M. Van Strydonck, J. Sergant, M. Boudin & M. Bats (eds.), *Chronology and Evolution within the Mesolithic of North-West Europe*: 615-634. Cambridge: Cambridge Scholars Publishing.

A PowerPoint presentation by Scott Timpany with various pictures of submerged forest deposits is available here:
https://www.academia.edu/10389858/Looking_for_lost_lands_-_submerged_forests_in_the_UK?email_work_card=title

Doggerland
Related to the study of submerged forests in Britain is the study of the North Sea land-bridge, for which an entry point may be found in the project of Professor Bryony Coles:
https://humanities.exeter.ac.uk/archaeology/research/projects/title_89282_en.html

The: foregoing is a format adapted version of the webpage available on the author's website at:

https://www.third-millennium.co.uk/submerged-forests-britain-ireland

11   R. Kay Gresswell and the Irish Sea Coast

In his 1953 book *Sandy Shores in South West Lancashire*, Ronald Kay Gresswell summed-up his investigations into the deposits of the South Lancashire Plain between Liverpool and the River Ribble. He described a wave-cut notch in the flat coastal plain, which he termed the *Hillhouse Coast* after a prominent landmark. Westward of this ancient shoreline he described the ancient forest of oak and silver birch that lay beneath the onshore deposits and which also occur as submerged forest deposits beneath the modern beach line from Liverpool to the Ribble. He went on to link this ancient shoreline to deposits in the Fylde and Cumbria. In the pre-radiocarbon era he could only estimate that the wood was of Neolithic age, correlated with the so-called 25-foot beach found in other parts of Britain.

Gresswell estimated the shoreline of the eastern Irish Sea after melting of the Pleistocene ice-sheet (p15) and (in his own summing-up): "*...the eustatic depression of the sea-level, which remained uncompensated by any other factor, must have caused the coastline to lie a considerable distance westward of its present position*". He drew this early Holocene coastline in Fig.6 as reproduced here. This would locate the early post-glacial coast at the 22-fathom depth line (approx -40m depth) which showed the Isle of Man linked to Cumbria.

Across this emergent rectangular plain (as may be seen on his map) he discerned from navigation charts where the ancient continuation of the rivers Ribble and Mersey-Dee had once flowed into a much-reduced Irish Sea. He argued that east of the 10-fathom depth-line the drowned river valleys had been smoothed by the sediment from the modern rivers; and he argued from his onshore borings that the in-filling of the Ribble estuary could similarly be observed in the Lancashire deposits.

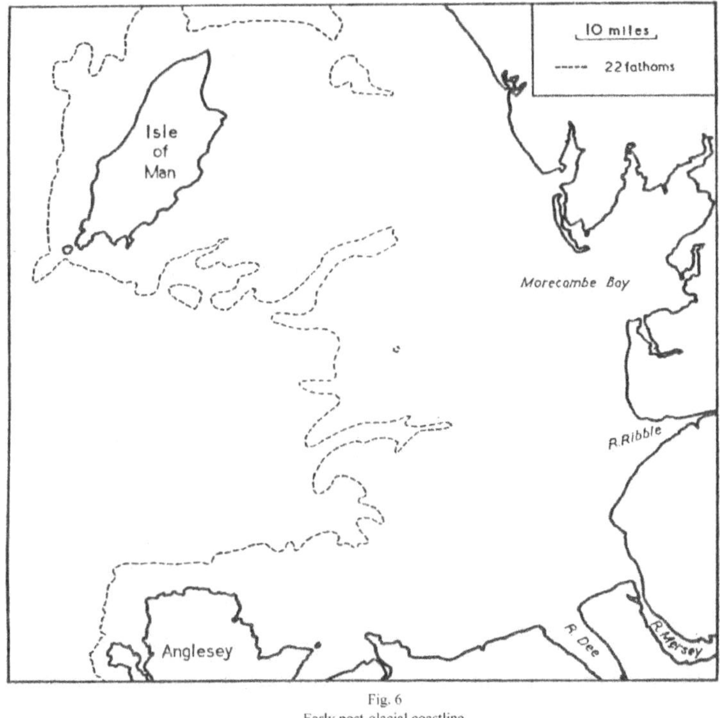

Fig. 6
Early post-glacial coastline

Figure 6 from Sandy Shores in South West Lancashire (1953)

In chapter IV Gresswell devised the sequence of advances and retreats of the post-glacial Lancashire coastline during the Holocene. On page 47 he illustrated this in the diagram reproduced here.

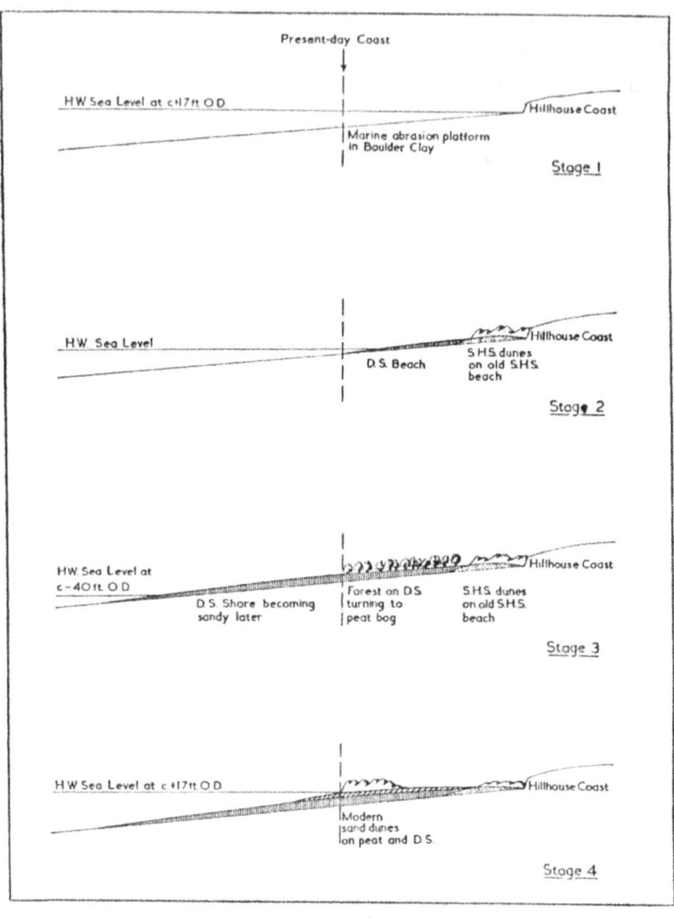

FIG. 13

Post-glacial episodes in South Lancashire

Diagrammatic and not to scale. S.H.S. = Shirdley Hill Sand, D.S. = Downholland Silt

Figure 13 from Sandy Shores in South West Lancashire (1953)

In stage 1 (early Holocene/ post-glacial) the sea rose to 7ft OD forming the Hillhouse coast. It then retreated in stage 2 towards the modern shore, allowing the growth of the oak and birch forests in stage 3 (mid-Holocene) as the sea retreated further out to -40ft OD. In stage 4 the sea returned close to modern shores, drowning the forests.

As a lifelong resident of Southport, Ronald Kay Gresswell M.A., PH.D., F.R.S.A., F.G.S. (1905-1969) was a prominent member of the British Institute of Geographers, known best for his fieldwork on the geomorphology of Lancashire during the 1940s and 50s. His work predated the radiocarbon and tree-ring dating methods that came to the fore only in the 1960s and 70s; and also the Milankovich theory of ice-ages that gained acceptance around the same time. His research is therefore 'pure' and unbiased by these later theories; he discusses the 'eustatic' variations of the post-glacial sea-level due to ice-melt and the recovery from the ice-burden, solely in terms of their apparent local effects.

As in all such studies of ancient sea-levels, local 'isostatic' vertical movements of the land can be conveniently conjured to make almost any theory fit the evidence – so long as one does not try too hard to correlate them with the parallels from other parts of the world. This is not a criticism of Gresswell for it is found in almost all such studies. Later researchers from the 1960s onwards, taking on board the then-prevailing theories of glacio-isostatic modelling, would revise Gresswell's theories on the grounds that they did not correspond with the wider picture derived from the worldwide sea-level curves and the melting of the Laurentide ice-sheet. A summary of this later research may be found in the Coastal Geomorphology Survey of Great Britain, chapter 7 (see the reference below, pp 4-5).

In the 1950s Gresswell could only date the period of lowered sea-level by the vegetation sequences. The consensus at that era, based on the earlier work of Godwin and Steers, was that the deciduous forests were of Boreal-Atlantic age. This view of a high sea level, followed by a regression below modern sea level and finally a rise to modern shores, was often described as "the submerged forest period". Radiocarbon dates can now help to date these forests more reliably; see the feature at chapter 10 here or online via:

Submerged Forests around Britain and Ireland

A stern critic of Gresswell's work has been M. J. Tooley who would argue (1978) that the Hillhouse 'cliff' was actually the shore of a trapped inland lake; and that the true former coast lay somewhat further west. From modern radiocarbon dates he and other researchers would now prefer a sequence of more conservative advances and retreats around the modern shoreline. Up to five possible transgressions of the present-day coast are envisaged between 9200 BP and 5000 BP; with subsequent build-up of the deep sandy beaches that we see today. The idea of a submerged forest period was quietly dropped. There remains general agreement however, that the sea reached its modern level around 5000 BP (c3000 BC).

It should be noted that none of this later work challenges Gresswell's insight that the former river channels could be discerned in the shallow bed of the eastern Irish Sea and therefore they must be post-glacial; and it follows that at whatever era these rivers flowed, the Isle of Man must have been linked to Cumbria. The divergence is that the later researchers, being constrained by the glacio-isostatic models that they must cite, could not contemplate such huge advances and retreats of the shoreline by as much as 30-40 km across the eastern Irish Sea; they would prefer to talk only of small-scale oscillations just offshore of the modern coasts. This is where Gresswell's naïve and 'pure' insight is so valuable.

In my own cross-disciplinary studies, I came at this problem from a world-wide pattern of correspondence of sea level change in alternate quarter-spheres. This is the pattern expected to be produced by a pole-shift that would have occurred during the mid-Holocene: the late fourth millennium BC (see the feature here: Raised Beaches and Submerged Forests). This theory was first published in *The Atlantis Researches* (1995) and *Atlantis of the West* (2002). We find emergence around 5,000 years ago, by similar distances, along shallow East Asian and South American coasts; and so we should not be surprised by submergence of the scale proposed by Gresswell.

The present author would continue to have serious issues with current models of sea-level change since the ice age. The various vertical movements of land and sea-bed (or the collapsing 'fore-bulges' of more modern parlance) which are proposed to explain the sea-level anomalies around the world would themselves constitute a change to the Earth's figure; and must therefore trigger a wobble of the rotation axis and a pole-shift. This would in turn feed-back as variations of sea-level and climate very similar to those that the 'isostatic' movements are advanced to explain. Sea level researchers seldom mention geodesy and fail to consider the rotational dynamics of the planet.

The question of whether the floor of the eastern Irish Sea was exposed during the warm mid-Holocene period has wider consequences for archaeology. It is also related to the question of whether there was a land-bridge between Britain and Ireland, and to the Isle of Man; and how the various flora and fauna reached these islands after the ice age. It offers the possibility of Mesolithic and Neolithic archaeology submerged offshore, compliant with Welsh, Irish and Mediterranean myths. Celtic myths and legends recall sunken cities and lost lands around the coast of Britain; and if we are to properly investigate these as memories of real events then the 'submerged forest period' prior to 5,000 years ago is the only era when we find both submergence and a warm climate.

In the BGS Report (2015) we may see that Gresswell's 10-fathom line would roughly correspond to the *Eastern Irish Sea Mud Belt* and the older region as the *Central Irish Sea Gravel Belt*. In both these regions however, the report describes the Holocene deposits as varying between 5m and up to 40m depth. Therefore anything of archaeological interest that was formerly at sea level would now be deeply buried. See: Patterns on the Irish Sea Floor.

In closing this review of Gresswell's pioneering work, I can only express appreciation for the painstaking work that he and other fieldworkers undertake; those who excavate in difficult locations to give us the dated primary evidence of past regimes;

and without whose work the secondary and cross-disciplinary researchers could not progress. But specialists can sometimes be too close to a problem; what we need is not more onshore work and theoretical models, but more data from offshore.

The diagrams reproduced here from 'Sandy Shores in South West Lancashire' are believed to be out of copyright. Should anyone object to their inclusion here then please contact the author.

References

The following will lead to all the relevant detailed coastal studies and papers.

http://archive.jncc.gov.uk/pdf/GCRDB/GCRsiteaccount1961.pdf

this is part of:
Geological Conservation Review
Volume 28: Coastal Geomorphology of Great Britain
Chapter 7: Sandy beaches and dunes – GCR site reports
Site: AINSDALE (GCR ID: 1961)

Gresswell, R.K. (1953b) Sandy Shores in South Lancashire: the Geomorphology of South-West Lancashire, *Liverpool Studies in Geography*, Liverpool University Press, Liverpool

Tooley, M.J. (1978) *Sea-level Changes in North West England during the Flandrian Stage*, Clarendon Press, Oxford

Mellett, C., Long, D., Carter, G., Chiverell, R and Van Landeghem, K (2015) Geology of the seabed and shallow subsurface: *The Irish Sea. British Geological Survey Commissioned Report*, CR/15/057.

This article was originally published as an interactive web-page in 2019 at:

https://www.third-millennium.co.uk/the-irish-sea-coast

12   Patterns on the Irish Sea Floor

Ask a geologist to do a survey of the sea bed and they will find geology. Any mark or feature in the surface sediments will be interpreted in terms of the geomorphology that they expect to find. An archaeologist looking at the same buried sub-sea features might see something different, as with crop marks on land. But why would an archaeologist wish to investigate the sea bed other than to seek, perhaps, an ancient shipwreck?

The possibility of ancient human settlement on what is now the floor of the Irish Sea – or indeed anywhere around the British or Irish continental shelf – is raised by the stories in Celtic mythology, which variously suggest lost cities around the coast: Lyonesse off Land's End, a submerged 'Celtic Otherworld', or a lost land called Annwn; and other myths from Mediterranean sources also pose questions. Welsh myths describe a plain "as large as the sea" off the coast of North Wales and Irish myths would portray a submerged "flowery plain" in the vicinity of the Isle of Man. Irish and Welsh legends speak of a golden tower in the sea explored by the earliest colonists. [1] To approach these myths logically either they are just that: 'myths' in the sense of 'fiction' or they must be based on an authentic root memory. If indeed they recall an ancient submergence then the next logical step is to ask, when?

We need not debate here the various mechanisms of sea-level change. On all conventional theories it must be remembered that the gradual eustatic rise of the sea since the ice age extended over thousands of years; a period of time longer than that which separates us from Dynastic Egypt. It is a failure of imagination to suggest that during these millennia there was no human settlement on the now-submerged regions of the continental shelf. The consensus is that present-day sea-level only stabilized around 5,000 years ago since which time there have been just minor fluctuations. In other parts of the world we find the opposite phenomenon at this same era: emergence and raised beaches rather than submerged features, together with evidence that people quickly colonised the newly exposed land (see: Raised Beaches and Submerged Forests).

For Britain and Ireland, submerged forest deposits are found all around the modern coasts (see: Submerged Forests around Britain and Ireland). These come from various ages, but most date from either of two periods; the first roughly 5-6000 BC around the time that the North Sea land bridge was flooded, while those around western coasts cluster around a later period closer to 3000 BC. So if we should seek any submerged archaeology then, logically, it should date from these eras. The earlier dates correspond to the Mesolithic period of supposed hunter-gatherers; the later dates correspond to the transition between the Middle and the Late Neolithic. Again, logically, any memories of real events in a legend would suggest that the former date is too early; and so we should be looking around 5,000 years ago to find underwater archaeology to match the legends.

We should expect that any submerged Middle Neolithic sites would look like those that we find on land from this era; such as the Court Cairns found all round the Irish Sea or the dolmens found further south; or perhaps settlements like those at Skara Brae; or field-walls such as those beneath Irish peat bogs. Cornish legends about the loss of a land called Lyonesse speak of some 140 'churches' drowned in a sudden rise of the sea. [2] We would have to consider these as pagan religious sites and parishes. What might such submerged cairns and settlements look like now, ruined and buried beneath metres of silt on the sea floor? Imagination is required. And this is before we even consider looking for 'lost cities'.

Older studies of sea level change around the British coast would speak freely of a 'submerged forest period' in the mid-Holocene (see: The Irish Sea Coast). The theory, before glacio-eustatic modelling of world sea-levels forced it out of favour, was that there was an initial period of eustatic sea level rise (i.e. melting of polar ice) which gradually drowned the coasts up to c.5500 BC. After this rise a regression exposed western coasts again, before a return of the sea to modern shores created the second wave of submerged forests around 3000 BC. Disagreement then arises as to how far out the sea may have receded during the warm mid-Holocene climate. Recent specialist studies would prefer a conservative 'somewhere west' of the present shores (as they have no data from the sea-floor and decline to speculate) or that the submerged forests were created by trapped freshwater lakes near the coasts.

The commonsense approach of archaeologists would be to seek submerged coastal sites such as we find at Westward Ho, Devon and at Skara Brae. However for the Irish Sea the legends of submerged plains, lost lands and towers are clearly pointing us further away from the shoreline than any archaeologist would consider; into depths that require expensive expeditions that so-far have been the preserve of geologists seeking oil and gas reserves or shallow sites for wind-farms.

Much of the geological mapping that we have for the central Irish Sea is acknowledged to be the work of just one man: the late Dr Robin Wingfield of the British Geological Survey (BGS) who, in the 1980s, identified ice-wedge polygons and circular features in sonar reflections taken north of Anglesey. [3] In the Strategic Environment Assessment (SEA) coastal resources survey of 2004 the Irish Sea region is defined as SEA6 and various reports are available via the BGS Geoindex database. Another survey investigated suitable sites for wind farms. This concentrated on an area closer to the Lancashire coast where the Morecambe Bay gas field and wind farms are situated and it did not encounter the patterned ground found in the older surveys; the sea-bed is instead described as mostly flat and featureless. The authors noted that the older sonar techniques were largely obsolete and that geophysical data was completely absent for large areas. [4] These results and subsequent studies are now embodied in the BGS Geoindex and in the updated report by Mellet-et-al of 2015, which summarises the state of knowledge as at that date.

The Geoindex shows the Holocene mud and sand deposits to vary between 5m and 40m thickness, with the deepest deposits lying off the Cumbrian coast. [5] Shipwrecks and modern pipelines are clearly identifiable from trailing sand patterns formed by the prevailing currents. Also noted are sub-surface features and 'pock-marks'; once those defined as escaping methane (marsh gas) are excluded there still remain some interesting sites to explore. Of particular interest is a feature lying off Morecambe Bay described as a pingo (on land we might call these kettle-holes). This reveals a rounded-rectangular feature, which would be a football-field-sized; and were it on land it would be about the same size as a typical Neolithic henge. Some of the pock-marks nearby are described as unknown and 'possibly man-made'. [6] This location would at some prehistoric era have been a river valley flowing into a lake that is now the Lune Deep. Another potential pingo is noted to the north of the Isle of Man.

To summarise the surface geology of the eastern Irish Sea as it is described would be as follows. The rectangular shelf slopes gently away from the English coastal dunes out to a depth of 50m between Anglesey and the Isle of Man. Close to the coast lies the Eastern Irish Sea Mud Belt overlaying the Central Irish Sea Gravel Belt that becomes exposed further west; this is mostly comprised of diamicton (a glacial till of mud and boulders). The smooth surface gives way to gravelly rough ground as the sea bed becomes deeper between Mann and Anglesey, where areas of bedrock are exposed. These rocks are a continuation of the onshore geology of Anglesey. Here the sea-bed has just a thin covering of Holocene mud and sand and it is in this region that the glacially patterned ground and polygons are described. This description may be investigated on the maps in the BGS database and its various options as linked here; together with the interpretive reports that have been published by the geologists. A summary map of the areas of 'patterned ground' and the other unexplained features is offered below as an introduction, since the BGS geological overlays give little indication of depth. Sea bed contours can be investigated via UK Admiralty chart 1826, or via the Admiralty online database.

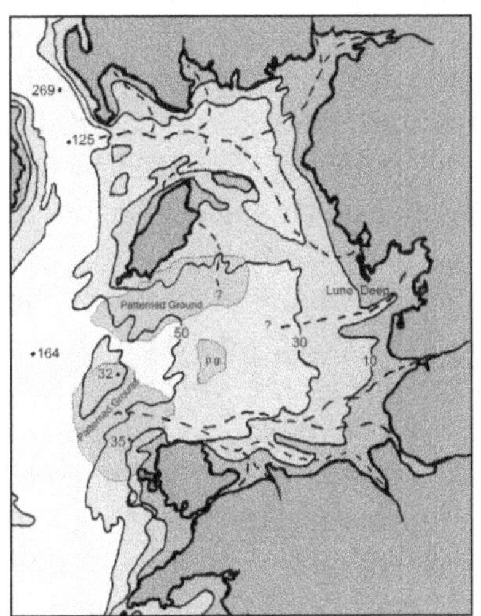

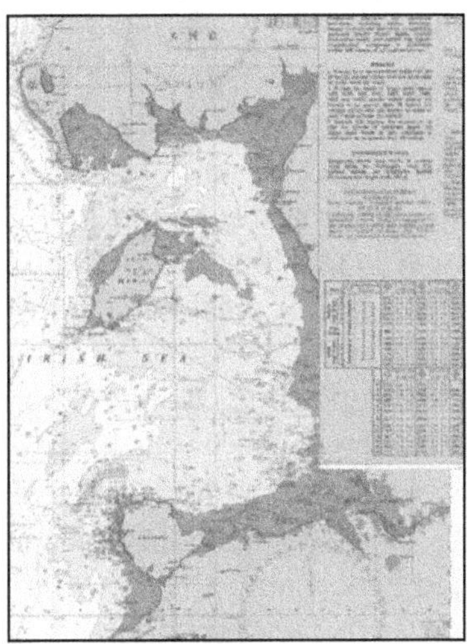

The public domain database of the British Geological Survey (BGS) can be found at:

http://www.bgs.ac.uk/GeoIndex/offshore.htm
and click on 'view the offshore Geoindex'.

The Admiralty wrecks and pipelines database plus a contour map (password required) is at:

https://data.admiralty.co.uk/portal/apps/sites/#/marine-data-portal

Geologists would deduce all of the patterned-ground to be glacial scour features similar to those studied on land. In his 1987 paper, Wingfield had interpreted the polygons and near-circular features north-west of Anglesey as collapsed pingos, which typically occur in formerly-glaciated areas. The small undulations on the sea-bed are interpreted as drumlins and flutes. It is important to note that there has been little examination of these features other than by sonar reflection and whatever they are formed-of lies buried beneath the sand and gravel. Another discovery from the DTI surveys is that the only place where hard bed-rock penetrates the Holocene deposits lies to the north-west of Anglesey and it is interesting that the two highest peaks on the sea-bed at 32m and 35m depth correspond to areas where the bedrock is exposed. [7] The Holocene sand and mud varies in depth as shaped by the prevailing currents and it is likely that much has been scoured away in this region.

In 2002 I was given permission to reproduce some of the sonar interpretations in *Atlantis of the West*. I do recall in the 1990s speaking by telephone to a representative at the BGS (who may have been the late Dr Wingfield) who told me there was no other evidence then available. Later, in 2002, I spoke to his colleague Ceiri James who informed me that Dr Wingfield had died; and that he alone would have known on what evidence he described the patterned ground (as published in 1995 and cited in the later SEA6 reports). Recent enquiries of BGS could not identify from precisely which surveys the published evidence of 'polygons', in areas other than those near Anglesey, had actually originated. The public database shows numerous sonar tracks crossing these areas.

However, it is not really the relict features from the ice age that should interest us; rather we should ask: *what lies hidden beneath the sand and mud layers?* Another interesting feature from the DTI Environmental Assessment was a straight north-south 'wall' about 3m (10-feet) high where the mud has piled-up against it. [8] This lies close to the mapped boundary between the mud and diamicton. The survey describes the straight feature as unidentified and probably man-made – as it was not at that time shown on the Admiralty databases of wrecks or pipelines. However, it does correspond to where the sonar track crosses the modern pipeline between the Morecambe Bay gas platform and its outlying northern well! This shows us how quickly the modern currents can pile-up sediment over even a recent man-made structure; so who knows how much archaeology might lie buried deeper beneath the Holocene mud and gravel.

In this analysis I would seek only to open minds to what is possible in order to reconcile the incompatible cross-disciplinary picture that arises when the same evidence is approached from contradictory specialist viewpoints: geology, archaeology and mythology, each offering us a different perspective. Archaeologists are only now waking-up to the likelihood of Mesolithic human activity on the submerged 'Doggerland' of the North Sea (Sunday Times 8-Sept-2019), so perhaps the Irish Sea is worthy of equal consideration:

https://www.thetimes.co.uk/article/doggerland-the-fertile-paradise-that-joined-uk-to-europe-emerges-from-sea-bed-t5t389ktl

I was reminded by all this of the story recorded by the Byzantine historian Procopius of Caesarea. In the fifth century AD he recorded the legend of a long wall built by the people of ancient Brittia (northern Britain) to separate the land of the living from the land of the dead to the west. [9] To this same place, we are told, the souls of the dead were ferried! Were a man or any wild creature to venture beyond the wall they would surely be struck down dead within half an hour by the foul air. This would be quite a fair description of dense methane gas escaping from muddy pools. After all swamp gas, we are told, can also catch fire and create UFOs to confuse the credulous observer; so perhaps we should not be taken aback by a little bit of potentially explosive archaeology on the sea bed!

Further Notes

1) For a discussion of all these legends see <u>Towers of Atlantis</u>. The legends of a submerged tower in the Irish Sea are found in Nennius and in the Irish Book of Invasions.
2) As described by William of Worcester (fifteenth century) in his Itinerary of Cornwall.
3) See Wingfield 1987 and later citations.
4) See page 52 of the Kenyon & Cooper Sandwindfarms report.
5) Various overlay-maps are publicly available online via the BGS Geoindex offshore website, of which many are reproduced in the Mellet-et-al report of 2015.
6) See figure 16 on page 33 of the Holmes and Tappin 2005 report.
7) See the summary map 'static bedforms' as Figure 13 on page 28 of Holmes and Tappin 2005, which lists the sources from which it was compiled.
8) See Figure 17.5 on page 34 of the Holmes and Tappin 2005 report.
9) Procopius, History of the Wars, VIII, xx

References and Sources:

Wingfield, R T R. (1987) Giant sand waves and relict periglacial features on the sea bed west of Anglesey. Proceedings of the Geologists' Association 98, 400–404.
https://www.sciencedirect.com/science/article/abs/pii/S0016787887800809

James, J.W.C. and Wingfield, R.T.R. (1987) Aspects of the sea bed sediments in the southern Irish Sea. Proceedings of the Geologists' Association 98, 404-406.
https://www.sciencedirect.com/science/article/abs/pii/S0016787887800810

D.I., Jackson & R.T., Wingfield & D, Evans & R.P., Barnes & M, Arthur & M, Howells & R, Hughes & Petterson, Michael. (1995) The Geology of the Irish Sea, London: HMSO for the British Geological Survey.

2004 Strategic Environmental Assessment SEA6 SV Lia SEA6 Survey - seabed sampling survey with video and photography (Irish Sea)
https://portal.medin.org.uk/portal/start.php?tpc=004_aba64100-c149-4de3-e044-0003ba6f30bd&step=0014

Kenyon, N. and Cooper, W. (2005) Sand banks, sand transport and offshore wind farms. 10.13140/RG.2.1.1593.4807.
https://www.researchgate.net/publication/285584613_Sand_banks_san d_transport_and_offshore_wind_farms

Holmes, R. and Tappin, D. R. (2005) DTI Strategic Environmental Assessment Area 6, Irish Sea, seabed and surficial geology and processes, British Geological Survey Commissioned Report, CR/05/057.
https://assets.publishing.service.gov.uk/government/uploads/system/upl oads/attachment_data/file/194659/SEA_6_Section_5_web.pdf
http://nora.nerc.ac.uk/id/eprint/11259/

Mellett, C. Long, D. Carter, G. Chiverell, R. and Van Landeghem, K. (2015) Geology of the seabed and shallow subsurface: The Irish Sea. British Geological Survey Commissioned Report, CR/15/057.
http://nora.nerc.ac.uk/id/eprint/512352/1/BGS_Report_Irish_Sea_Geol ogy_CR-15-057N.pdf

This article was originally published as an interactive web-page in 2019 at:

https://www.third-millennium.co.uk/patterns-on-the-irish-sea-floor

13   Fifteen Years On From Atlantis (of the West)

A short summary article originally published in SIS Review in 2017

In 2003, my book, *Atlantis of the West – the Case for Britain's Drowned Megalithic Civilization*, was published. It was actually a second edition of *The Atlantis Researches*, published in 1995. The new name was chosen by the publishers as they thought it might sell more copies in UK and USA, but the original subtitle, "The Earth's Rotation in Mythology and Prehistory", gives a better idea of what it was really about. In 2005, my follow-up, *Under Ancient Skies*, was published [1]. This is perhaps less well-known than its predecessors but, in many ways I was happier with its content. In the same year, I gave a talk at the SIS Autumn meeting on the subject of catastrophism around 3100 BC, its causes and effects [2].

I offer this short review article to summarise the theories because, although in the years since publication I have done very little new work, the ideas in the books keep appearing in various places - some by people independently coming to similar conclusions; others citing my own ideas third-hand without knowing from where they originated. I had hoped that my books might move the debate forward from the 1950s science of Velikovsky, Hapgood and others, onto more modern evidence – yet still I see books and TV series about catastrophist subjects that seem to be little more than marketing concepts to profit from mystery themes. My work has been out-there a long time, so any author who does not cite it now is guilty of poor research, even if all they wish to do is to debunk it.

So, how might I summarise the original theories without missing something vital? Not easy, because true research is complex. It isn't just a case of pointing to three stars in Orion's belt and then padding it out to make a book. So let me begin by highlighting the case for a mid-Neolithic crisis around 3100 BC, for which there is a growing cache of evidence from many disciplines. Some other investigators prefer a date of 3200 BC but we mean the same thing [3]. From both scientific and ancient sources we may see a convergence here:

Atlantic/Sub-Boreal climate changes	3300-2900 BC
Sea-level changes worldwide	3200-3000 BC
Greenland high-acidity event	3250 BC
Tree-ring low-growth event	3199 BC
Mesopotamian/ Biblical Flood	3100-2800 BC
Indian Calendar Era (Kaliyuga)	3102 BC
Mayan Calendar Era	3113 BC
Egyptian First Dynasty	3110 BC
Sahara Desert climate changes	3000 BC?
Radiocarbon tree-ring date corrections	1000-3000 BC?

I add Plato's story of Atlantis to this list because it is an account, presented as history, of a catastrophic submergence around the beginning of the Egyptian state. This would place it at somewhere before 3100 BC; not at 11,000 years ago – a fabulous antiquity which arose only because the Egyptian priest included a dynasty of long-lived demigods before the dynasties of real kings (as is found on the Palermo Stone). This is one example of what I called a "mythological fossil" – a detail preserved within a legend that can be taken out and separately analysed using quantitative science.

Another example was that very preqise statement in Plato's story that the ancient kings gathered "every fifth and every sixth year alternately" (not, you will note, a vague "every five or six years"). This suggested to me that they were using a calendar;

so I set out to see if such a calendar would be accurate. In *Under Ancient Skies* I showed that it indeed does work. There is a pre-Roman Gaulish calendar called The Calendar of Coligny that appears to academic analysis to be a quite hopeless attempt by the Druids at a 5-year lunisolar cycle. However, when you combine it with an alternating 6-year cycle then it becomes potentially more accurate than the Gregorian calendar. It was this detail which convinced me that Druid astronomy must have been a survival from the megalithic society of Atlantic Europe. Once you are open to this possibility, then, many other pieces of evidence drop into place to support it, alas too many to go into here. So how much more of the story of a mid-Neolithic catastrophe might eventually prove to be true?

Another way to examine a Neolithic flood event is to look for the geophysical evidence that it must leave behind. Whatever simplistic mechanism might be proposed to cause 'the great flood' or to sink a large landmass then it must have worldwide effects that result from the fact that the Earth rotates on its axis. If land and sea move, then the Earth's rotation would wobble and the poles would migrate. The mathematics that describes this motion is well known to specialist geophysicists and astronomers. My assertion was that such numerate academics set little value upon catastrophist theories or upon myths as evidence, and they do not recognise the evidence that is before them.

The phenomena known as the Chandler Wobble and the nearly-diurnal wobble, if excited by sufficient force, would give rise to seven-year rhythms in the climate. This interval occurs in many ancient sources and is another of those mythological fossils that we can examine. The most obvious example of course is the Biblical Joseph story. Although many people have written to me over the years about the various theories in my books, not one has picked up on this. I thought some might seize on it as proof of the Bible and such, but surprisingly, no.

These climate variations would arise because the 14-month Chandler wobble as modified by the other mode would combine with the 12-month seasons to give 7-year climate rhythms that decay after about 20 years; a very simplistic summary, since there are other longer term effects, such as changes in the figure of the earth (the geoid). A pole shift would cause a pattern of long-term land and sea-level changes. My own theory suggested submergence along the Atlantic coasts and apparent rising of land further east. Incidentally, one piece of evidence that I did not pursue was the apparent uplift of land in central Asia since about 3000 BC that has resulted in the shrinking of the Caspian and Aral Seas and the rivers of that region as described by ancient writers – a book theme itself for someone.

I also touched upon whether these changes in the Earth's rotation might cause a permanent shift in both the axis-tilt (obliquity) and the length of the day, in addition to a geographical pole shift. Although I presented the evidence, I remain less convinced of this aspect. The problem is that while a pole shift could arise by movements within the Earth, changes to the axis tilt and diurnal period require a quite enormous external force to be applied. It is this which makes conventional academic opinion so sceptical and rightly so.

Nevertheless, if we look at the construction of the Mayan, Chinese and Indian calendars they show a belief in a former year of 360-days and cycles of 60 and 432 days that would be needed to track a real nutation of the axis. Another pointer is the calibration of radiocarbon dates that becomes necessary before 1000 BC to bring them into line with tree ring dates. This suggests a varying magnetic field after about 3000 BC; but why should this be so? The magnetic field is generated by the daily rotation of the Earth's iron core; and why should this have changed? My theory was that a nutation of the axis (the nearly-diurnal or core wobble) persisted through the next two millennia* and explains the need for the additional calendar devices. It also offers a reason why Neolithic monuments in the third millennium BC appear to be astronomically aligned.

One last thread in my research was that by looking at references to solar and lunar eclipses in ancient sources and the retro-calculated variation in delta-T then any postulated impact disturbance could not have occurred since the second millennium BC; and that the modern solstice alignments of Neolithic monuments (Stonehenge, Newgrange, etc) would not be valid if the causal event occurred any later than about 3100 BC.

Whatever may have caused the mid-Neolithic crisis could have been no ordinary comet or asteroid impact or else its effects would be obvious. Direct evidence for an impact event has yet to be found; there is no 'crater'. A solar comet of the mass required to affect the Earth's rotation would have to be so large that it should have caused a mass extinction. So my conclusion was that it must have been something very small and travelling very fast; perhaps a piece of supernova debris; or some phenomenon of new physics that we do not yet understand.

Without direct evidence, the case for a mid-Neolithic event requires the appreciation of a pattern of evidence across many disciplines [4]. Single-subject experts do not see this and do not pursue it. It is all too easy to dismiss such evidence without proper scrutiny, even though it bears no relationship to the 1950s ideas that sound like just so-much phlogiston to modern ears. Minds may be more open to catastrophism now than twenty years ago but there remains a long way to go.

Notes and References

1. Paul Dunbavin, *The Atlantis Researches: the Earth's Rotation in Mythology and Prehistory*, Third Millennium, 1995; *Atlantis of the West: The Case for Britain's Drowned Megalithic Civilization*, Constable & Robinson (Carroll & Graf in USA), 2003; *Under Ancient Skies: Ancient Astronomy & Terrestrial Catastrophism*, Third Millennium, 2005.

2. Report on Autumn Meeting, 2005, C&C Workshop 2005:3, p. 2. The following unpublished papers were intended for SIS: P. Dunbavin, *'Updating George F. Dowell'*, 2006; *'Plato, the Neolithic Calendar and the Evidence for Catastrophism'*, 2006; *'Akhenaten, Eclipses and the Chronology of the Egyptian XVIII Dynasty'*, 2006. **

3. A range of useful catastrophist books and research papers, including the above, is listed by William I. Thompson in Celestial Catastrophism - Bibliography and Handbook.
http://www.creationism.org/books/BibliographyCelestialCatastrophism.htm.

4. http://www.sis-group.org.uk/news/piora-oscillation.htm

***Errata:** the word 'decades' that appeared in the original journal has here been corrected to 'millennia'. The Chandler wobble lasts for decades; the core-wobble, once triggered would persist much longer.

****Note added in 2020**: All of the above books, articles and research papers are now available, along with many others via the author's website at www.third-millennium.co.uk

Citation: *Chronology & Catastrophism REVIEW* **2017:3   pp53-4**

14   Submerged Islands opposite the Strait of Gibraltar

Plato's dialogues give us a clear indication of where he believed the sunken island of Atlantis to have been situated. However, modern commentators, academic and popular alike, tend to disregard his actual words. This has led them in different directions: to seek alternative locations within the Mediterranean and elsewhere. So firstly, let's examine what he actually says.

...in those days the Atlantic was navigable; and there was an island situated in front of the straits which are by you called the Pillars of Heracles; the island was larger than Libya and Asia put together, and was the way to other islands, and from these you might pass to the whole of the opposite continent, for this sea which is within the Straits of Heracles is only a harbour...
[Timaeus, Jowett translation]

In the Critias Plato describes a peninsula of Atlantis: *"facing the country which is now called the region of Gades [Cadiz] in that part of the world"*. Therefore it is clear that Solon's Egyptian sources are describing a location outside the Strait of Gibraltar.

In 1979 a team of Russian investigators led by Andrei Aksyonov would claim to have found walls and staircases on underwater photographs taken on the Ampere Seamount, which lies some 450 miles west of the Strait of Gibraltar. This is a prominent feature on navigation charts peaking at 56 m depth, together with the Gettysburg Seamount at just 35 m depth and nearby Ormond Seamount at 23 m depth. The Russian reports have been subsequently pursued by various authors as the site of a 'sunken continent', or of an island colony of Atlantis. The details of the Russian and related claims have been amply recorded on the excellent Atlantipedia website and therefore need not be repeated here in such detail; for numerous links on this subject see:

http://atlantipedia.ie/samples/tag/gettysburg-seamount/

We should perhaps be wary of Russian research. A search for Atlantis would have made excellent cover for Cold War underwater spying activities, after the manner of the U.S. Glomar Challenger expedition and the search for the lost Titanic! Aksyonov would later declare that all the submerged features were entirely natural. After 1986 all went quiet.

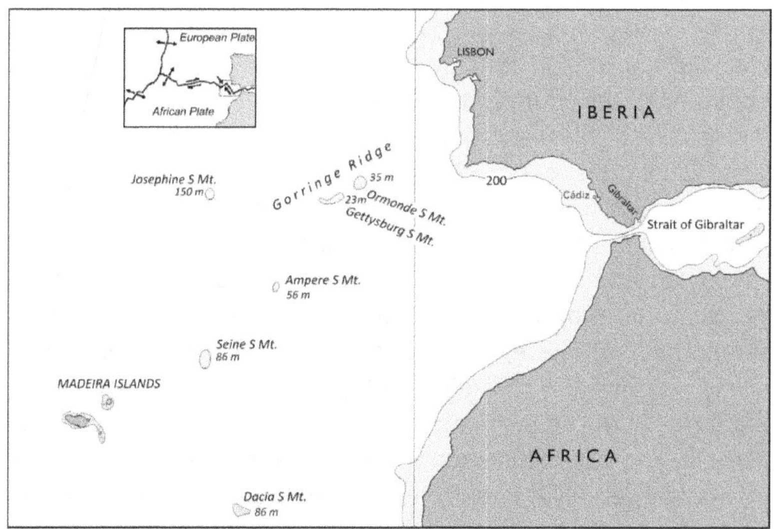

A summarised bathymetric chart showing coasts, islands and seamounts above 200m depth, lying opposite the Strait of Gibraltar. The transform fault passes through the Gettysburg Seamount.

The geography of these submerged seamounts is now much better known and the bathymetry of this region has been explored to a level of detail that we could only wish for most other undersea locations The Ormond and Gettysburg Seamounts lie on the Gorringe Ridge which runs along the Azores-Gibraltar fracture zone, where the African Plate is colliding with the European Plate and pushing up the Iberian peninsula. A colourful introduction to the oceanography of the Gorringe Ridge may be found at:

https://eu.oceana.org/sites/default/files/reports/seamounts_gorringe_bank_eng.pdf

A detailed bathymetric study of the seamounts on the Gorringe Bank is that of de Aletriis et al, which records the cruise-expedition that surveyed the entire area in 2003. [1] The mapping reveals the two seamounts as convex, almost flat, 'shelf' areas, which they describe as resembling 'Pacific-type' guyots (submerged coral atolls).

The studies published by the various oceanographers will give the impression that the two seamounts could have been above the sea during the lowered sea-levels of the late-Pleistocene. However, in the 2003 survey the authors admit that: *"dating these features is still not allowed by present data"* – a typical plea for more data that is almost standard in sea-level studies! The tentative dating comes solely from the usual glacio-isostatic modelling assumptions. Terraces are found at depths of 60m on the Gettysburg slopes and at 120m on both seamounts. [3] These correspond with terraces of similar depth found along the Spanish-Portuguese coasts. Four main platforms have been recorded, at depths of 35–40m, 60–70m 80–90m and 120–140m; some being tilted by about one degree. [5] As the sea level is believed to have risen by around 120m since the Ice Age maximum then it is assumed that any such deep features must therefore be of glacial age. Other papers then cite this and thus it enters the chain of evidence as unquestioned fact.

Portuguese researchers have researched the history of earthquakes and tsunamis in this region in pursuit of the causes of the great Lisbon tsunami of 1755. The earliest historically known quakes were recorded by Medieval Spanish and Portuguese chroniclers at 246 BC, 218 BC and 210 BC. [2] These tsunamis may be associated with the disappearance of the lost city of Tartessos somewhere in the Gulf of Cadiz, which some have compared with Atlantis. [3] The earthquakes have their origin along the Azores-Gibraltar fracture zone; the Gorringe region being a seismically complex zone where other faults converge and where subduction can occur. The Spanish coastal sediments record many earlier tsunamis – but this can tell us little about stable ancient coastlines, as tsunami waves can penetrate far inland. [3]

Scientific research places little value on legendary evidence, such as that of the classical historian Diodorus Siculus. He records that a low-lying region called the Marsh Tritonis lay "near the ocean which surrounds the earth" and that it disappeared in the course of an earthquake "when those parts...which lay towards the ocean were torn asunder". [4] The earthquakes are discussed by Diodorus along with events during the earliest period of the Egyptian state (predynastic or First Dynasty c.3100 BC). Therefore they may remember those same earthquakes that Solon attributed to the Mediterranean region as far away as Greece and occurring along the same fault line, as it continues east from the Straits to Malta and Greece.

Unlike volcanic Madeira and the Canary Islands further out, the two Gorringe seamounts could only have been flat-top atolls; more like modern Bermuda. It should be apparent that even if these two islets did appear above the sea during the Holocene then they cannot have been Atlantis. They are simply too small to hold all the features described.

The modern bathymetric studies cited above are highly detailed and they offer no evidence of patterned ground or any obvious human structures. They offer nothing to support the Russian claims of the 1970s to have seen walls and buildings.

However, the submerged seamounts could explain the memory of an island lying 'opposite Gades' or opposite the 'Pillars of Heracles', which was the way to other islands beyond. The seamounts could have been shallow islets until 10,000 years ago (Plato's Atlantis date) or even as recently as 5,000 years ago when there is general consensus that the 'post-glacial sea-level rise' finally stabilized.

As an author on these subjects, I can scarcely argue for a lower sea-level during Neolithic times five-thousand years ago, without at the same time acknowledging that small islands may have existed on the Gorringe Ridge. I did note this possibility briefly in *Atlantis of the West,* which predated the above Spanish and Portuguese studies; but as with the submergences around Malta and Sicily, there was no field evidence to cite at that time. Perhaps it is now worth someone taking another look for archaeology and dating evidence on these submerged terraces?

References

1.Alteriis, G.., Passaro, S. & Tonielli, New, high resolution swath bathymetry of Gettysburg and Ormonde Seamounts (Gorringe Bank, eastern Atlantic) and first geological results, Mar Geophys Res (2003) 24: 223.
https://doi.org/10.1007/s11001-004-5884-2.

2. Álvarez Marti-Aguilar, M. (2017)
https://www.academia.edu/35530049/_Tsunamis_in_the_Iberian_Penin sula_during_antiquity_the_historical_sources_5th_International_Tsun ami_Field_Symposium_Lisboa_3-5_September_2017

3. Abril, J-M., Periáñez, R. and Escacena, J-L. Modeling tides and tsunami propagation in the former Gulf of Tartessos, as a tool for Archaeological Science Journal of Archaeological Science, Vol 40, Issue 12 (2013) pp 4499-4508

4. Diodorus Siculus, Histories, III, 53-55

5. Pastouret, L., Auzende, J-M., Le Lan, A. (1980) Temoins des variations glacio-eustatiques du niveau marin et des mouvements tectoniques sur le banc de Gorringe (Atlantique du Nord-Est) **Paleo, 32, (1980) 99-118**
https://www.sciencedirect.com/science/article/pii/0031018280900346?via%3Dihub

This article was originally published as an interactive web-page in 2019 at:

https://www.third-millennium.co.uk/submerged-islands-gibraltar-strait

15   Catastrophes from Atlantis to the Aegean

Summary: *this discussion compares the popular-academic rationalisation of Plato's Atlantis as a fictional memory of the Thera-Santorini eruption with the alternative based on a stricter adherence to the internal content of the narratives. It also posits that the ultimate source is Libyan-Egyptian rather than Greek and that any solution must also comply with the additional semi-historical information supplied by Diodorus Siculus in his Histories. This is taken together with an analysis of the known geological and climate events of prehistory as compared to those suggested in the narratives.*

Modern commentators who discuss Plato's Atlantis and its demise will typically fall into two camps. The first would be the scholars of Greek literature and others that follow their lead, who prefer to retain the story within the realm of Greek mythology; they would treat it as Plato's literary fiction based solely upon his own philosophy and knowledge of the ancient world. Typically such commentators rationalise Atlantis as a hazy memory of ancient Crete or of the Thera (Santorini) eruption and tsunami that destroyed the city of Akrotiri on that island. However, this supposition ignores many internal statements of the narratives and neglects the additional details about Atlantians given by Diodorus Siculus. The hypothesis has become a regular feature in various much-repeated television series linking Atlantis and ancient Cretan civilization.

The second group of commentators are the diverse popular enthusiasts, who would seek to locate the lost city in a variety of places, from the Caribbean to the poles, selecting only those elements from Plato that they need and discarding the rest. Many of these also neglect Diodorus Siculus.

The Athens Acropolis and Mount Lycabettus were occupied at least 6,000 years ago during the Greek Neolithic.
Picture from: www.athensgreecenow.com

There is a reasonable third approach: which is to consider Plato's account for the most-part truthful and to analyse Atlantis and its catastrophe as a degraded history of ancient events; preserved in Egypt, and brought back to Greece by the sage Solon in the sixth century BC just as Plato says. Any proposed solution must therefore respect all the details of the ancient story that we have inherited from Solon's Egyptian source: Sonchis of Saïs. It is then possible to compare the legend with Egyptian rather than Greek tradition; and with the wider tradition of North Africa as I did in *Towers of Atlantis* and earlier books. This would transfer tenure of the story away from Greek scholarship to the Egyptologists — but these specialist-archaeologists don't want it! They would shy away from being tainted by speculation about Atlantis.

Those who regard Atlantis as Plato's fiction seldom look beyond the Thera eruption of c.1625. For this to work they have to discard the fabulously long chronology supposedly given to

Solon by the Egyptian priest; indeed his estimate of 9,000 years before his own time merely reinforces their scepticism as there is scant evidence of settlements in Greece from such an early date. Plato's timescale therefore has to be reduced by a factor of ten to make it credible. Another argument employed is that Plato's references to bronze would mean that Atlantis could not be older than the Bronze Age. However, as I argued in *Towers of Atlantis* the date of the Bronze Age itself is constantly being pushed further back in time by archaeological discoveries and is now set as early as 3200 BC for the Aegean. Those who wish to pursue the historiography may like to read the classic paper of 1960 by A.G. Gallanopoulos, which discussed the original excavations of Professor Marinatos, along with a survey of ancient Greek tsunami references that retain their relevance.

https://www.sciencedirect.com/science/article/abs/pii/0025322784900227

The true dating indicators within Plato's narratives are the statements that Atlantis and its catastrophe fell before the establishment of the Temple of Neit and its institutions at Saïs in the Egyptian Delta. In *Timaeus* the priest puts the foundation of Saïs at 8,000 years before his own time and that of Athens 1,000 years earlier; in *Critias* he says that 9,000 years had elapsed. This association would place the events right at the foundation of Dynastic Egypt, around 3100 BC or during the earlier predynastic era – and therefore at least 1500 years earlier than the Thera eruption. Those unfamiliar with this Egyptian goddess may like to read:

https://www.ancient.eu/Neith/

https://www.ancient-origins.net/ancient-places-africa/pharaonic-royal-city-sais-leaves-few-clues-researchers-002352

Neit was a war goddess of probable Libyan origins equivalent to Greek Athene. The religion of Neit was at its strongest during the First Dynasty after the unification of Upper and Lower Egypt. Egyptologists find the *mastaba* tombs of the First Dynasty kings – and also their queens who bore titles derived from the goddess Neit. It follows logically that if the Temple of Neit was established in the delta as long ago as the First Dynasty then we cannot set the era of Atlantis, *or the catastrophe that destroyed it*, any later than this date. The long-chronology, stated by Sonchis to Solon to derive from "sacred records", can then be seen as entirely consistent with the long-reigns of gods and demi-gods whom the Egyptians believed had ruled the Nile valley prior to the dynasties of mortal kings. [see Note 1] It is therefore relevant to seek a geological catastrophe that could sink an island or a city from a time before the establishment of the First Dynasty – the late fourth millennium BC.

Another internal detail is that the submergence of the city and its island corresponded with great earthquakes in Greece that destroyed ancient Athens and its armies – who had just fought-off an attack by Atlantians from the west. We may ask: what links Greece and the Atlantic Ocean? The obvious connection is the geological fault that runs from the Azores in mid-Atlantic through the Strait of Gibraltar, onward to Malta and Sicily in the central Mediterranean and then to the Aegean Sea. The geologists tell us that this is where the African continental plate is being subducted beneath the vast Eurasian plate; and the Aegean micro-plate is all that remains of the ancient Tethys Ocean that separated them. Subduction zones and their volcanoes are the most unstable regions on the planet; where earthquakes tend to be at their greatest magnitude, but occur only after deceptively long intervals of apparent stability.

Before we examine the geology, we should also review the parallel chronicle offered by Diodorus Siculus, who also tells us about *Atlantians* (a slightly different spelling but clearly related) and who possessed great cities. He goes-on to describe a matriarchal race called the Amazons, who similarly worshipped the goddess Neit-Athene; they conquered an Atlantian city named Cernê situated somewhere along the Libyan (North African) coast. The city of the Libyan Amazons was called Cherronesus ('peninsula') and it lay in the marsh Tritonis; generally identified with low-lying Chotts of Tunisia, but no-one really knows where it was. At some later time this city too was submerged in a time of great earthquakes:

> *...the marsh Tritonis disappeared from sight in the course of an earthquake, when those parts of it which lay towards the ocean were torn asunder.* [See Note 2]

It is then logical to ask whether this remembers the same geological event that is recalled in the Atlantis legend. Plato, in *Timaeus* also says: *"later... [there came] earthquakes and floods of extraordinary violence"*. If we treat Atlantis as an Egyptian rather than a Greek history then we have here two independent accounts, both ultimately of north-African derivation, that recall: *Atlantians*, the goddess *Neit-Athene*, a catastrophic *submergence* and a great *earthquake*. Too many coincidences here to just ignore!

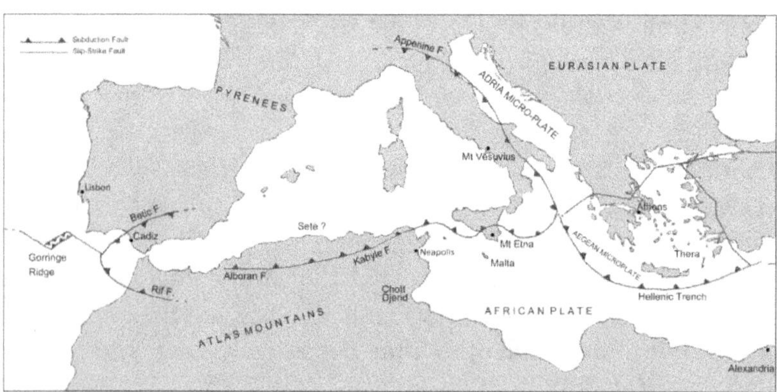

A tectonic puzzle! The Mediterranean Sea is a seismically complex region with numerous faults, some of which are ancient and inactive. Shown here, for simplicity, are the principal active subduction faults or crumple zones together with the normal faults where sideslip may occur. A mega-quake in one region could trigger other events further along this line of activity.

It is sometimes forgotten that we have very few ways of dating ancient earthquakes and volcanic eruptions unless there is a historical report. For example archaeologists still argue about the true date of the Thera eruption – some want to place it around 1560 BC; others following carbon-14 and tree-ring dating prefer 1600-1645 BC. For ancient earthquakes we can say even less; even if there is some artefact to be carbon-dated, the magnitude of an ancient quake can be only subjective specialist opinion. Some idea of the difficulties can be gained from the significant earthquakes database at:

https://www.ngdc.noaa.gov/nndc/struts/form?t=101650&s=1&d=1[1]

To attempt a dating of the earthquakes recalled by Diodorus is therefore problematic; the specialists will only trust a firm historical report, whereas the historical investigator wants the geologist to suggest a date. From the wording we can see that the ancient earthquake was no minor episode; Diodorus recalls that low-lying regions towards the Ocean were "torn asunder". This sounds very much like a mega-quake (magnitude 8.5 or greater) of the kind expected along a subduction zone, resembling those experienced more recently in Alaska and Chile. Diodorus Siculus or his lost source *could not have known* that two thousand years later geologists would discover that the Algerian coast was a mountain-building subduction zone, in order to invent a 'mythical' ancient catastrophe in just the right location. This kind of coincidence is what I have elsewhere termed a 'mythological fossil': an element within a myth that can be extracted and independently tested by science. Somewhere along the Mediterranean coast between Tunisia and Morocco there may be two lost cities waiting to be discovered, but precisely where cannot be discerned from the vague geography of Diodorus! It should be noted that neither of these lost cities would be Atlantis – that was somewhere else.

To commence at the western-end of the line of faults, in the Atlantic Ocean, we find the islands of the Azores; for so long naively claimed as the peaks of the sunken continent. We can now see these mid-ocean islands as a volcanic product of the transform fault that links the mid-Atlantic ridge of sea floor spreading to the subduction zone of Iberia-Africa. The nature of the fault therefore changes around the region of the Gorringe Ridge to the west of Cape St Vincent, where we find flat-topped seamounts that were formerly volcanic islands. Reliable evidence, of precisely when they were above the sea is still unavailable. Where oceanic plates of equal density collide they tend to push-up volcanic island chains and one day the ridge will become a new peninsula of Iberia.

North Africa

East of the Gorringe Bank the fault splits, with one active branch running north-east across the Sierra Nevada of southern Spain while the other continues through Morocco. Here we may hope to find more evidence of ancient earthquakes. Perhaps the various ancient legends surrounding the Pillars of Heracles do remember a series of real quakes that altered the coastline. The true active fault zone continues along the North African coast: through Morocco-Algeria-Tunisia. The largest documented earthquake here was that at magnitude-7, which occurred on 10 October 1980. [2] This is far from the worst to be expected of a subduction zone and those recorded further east in Tunisia tend to be of lesser magnitude. [3]

A catalogue of early historical earthquakes in Algeria lists one that occurred in Setif in 419 AD, magnitude unknown, but severe enough to be recorded. [4] Early historical quakes are recorded in Tunisia around 412 AD. A specialist dare not, of course, cite the legendary account of Diodorus Siculus or his lost source Dionysius as 'evidence'. Presumably Dionysius was citing historical sources in the Great Library of Alexandria. For comparison, Old Alexandria itself was lost to the sea following the earthquake and tsunami of 365 AD (estimated magnitude 8.5 off Crete). An account of this tsunami and its effects are given by the Roman historian Ammianus Marcellinus:

http://www.tertullian.org/fathers/ammianus_26_book26.htm#C9

The same tsunami may have been responsible for the inundation of Roman Neapolis in Tunisa. It is therefore not unreasonable to hold that older settlements have also been lost to ancient quakes further west along the coast of Africa.

Iberia

The Gorringe Ridge is believed to be the epicentre of the great Lisbon earthquake of 1755 that destroyed the colonial economy of Portugal, if not quite the city itself. Recent estimates rate the magnitude as no more than 8.5 – not quite the maximum 9.0 associated with a true subduction zone. An excellent account of this earthquake, leading to numerous references, is given at:

https://www.volcanocafe.org/the-lisbon-earthquake/

Plato quite clearly locates his lost island and its city in the Atlantic Ocean beyond the Strait of Gibraltar, but we shall not pursue that thread here. The evidence of Diodorus, from a completely independent source, would mitigate any suggestion that it was Plato alone who gave the Atlantic Ocean and its inhabitants a fictional name. Why has conventional scholarship neglected this corroborating evidence from Diodorus? Perhaps because he offers us mythology derived from unfashionable African sources, which attempts to rationalise the 'pure' Classical myths taught in schools and universities.

Elsewhere in his histories, Diodorus offers a rationalised summary of the legend of Heracles and his voyage to the west. This is Heracles the man, not yet the god! The hero's legendary voyage and labours *were a later event* than the war between the Atlantians and the Libyan Amazons. [Note 3] It is apparent from this narrative that Heracles set up his legendary pillars at what we now know as the Gibraltar Strait; and his voyages continued further west into the Atlantic, to Gades (modern Cadiz in Andalucía) and north into Gaul. This independent source contradicts any suggestion that Plato or Solon did not know where the legendary pillars were situated.

Plato and his source are consistent that the lost city was

situated outside the straits and this has led some commentators to link it with another legendary Bronze Age lost city named *Tartessos*; believed to lie beneath the low-lying marshland between Cadiz and Seville, where silver has been mined since ancient times.

https://www.ancient-origins.net/news-history-archaeology/2500-year-old-city-buried-under-flood-sediment-may-belong-lost-civilization-020521

A lost city here there may have been, but similar arguments may be advanced against this identification as apply for Thera and Akrotiri; It fails to meet the Atlantis description both in date and geography. The region is subject to the same geological influences as were discussed for the Lisbon earthquake. For an investigation of the earthquake threat in this region see:

https://www.agenciasinc.es/en/News/The-Andalusian-coast-has-suffered-a-large-tsunami-every-thousand-years

The onshore record suggests a major tsunami in the Gulf of Cadiz around 4,000 years ago, with perhaps as many as eight over the last 7,000 years. [5] We should also consider the effect of tsunamis created by underwater landslides. These may be triggered at any time by forces not necessarily associated with major earthquakes. In addition to the 1755 tsunami earlier underwater mud-flows (turbidites) are dated to 218 BC, 3010–3960 BC and 7765–8065 BC. [6]

It is conceivable that the legendary report of Diodorus recalls an underwater landslide event of the fourth millennium BC – but to accept an earthquake/tsunami solution here for the Atlantis 'quake' would require a mechanism to transmit its effects within the Mediterranean as far as Tunisia and Greece.

Malta and Sicily
Pivotal within the central Mediterranean is the position of Malta. Nothing from the Platonic tradition connects the islands

to Atlantis, but the central Mediterranean too has been suggested as a location of the sunken island. [7] It is easier to associate Malta with the tradition of Diodorus than of Plato. The Libyan Amazons originally came from an island too, identified as having a volcano. The only large island in this area that is home to a volcano is Sicily. We may consider whether Malta and Sicily were once linked, politically if not physically. It is also relevant that the temple-building period on Malta (c3500-2500 BC) predates the establishment of Dynastic Egypt.

The zone of faulting takes a loop offshore north of Tunisia, passing through Sicily and around the shores of the Adriatic Sea forming the Adria micro-plate. Although we find in Italy a zone of frequent earthquakes and volcanism, these are not the 'mega-quakes' that occur along subduction faults. Although locally destructive, they are not usually powerful enough to generate tsunamis or the widespread destruction that was associated with, for example, the Lisbon or Cretan earthquakes. In this region it is the volcanoes that present the true danger.

In a recent investigation of sea level changes around Malta by Furlani-et-al, the specialists conclude that the island has been 'stable' since the Ice Age and follows the predicted eustatic sea-level curves. They see no need to invoke local isostatic or tectonic movements that specialist researchers employ to rescue divergence from the expected eustasy. [8] The study cites the sea level curve of Lambeck-et-al (2011) for the Mediterranean Sea; indeed they conclude that this region has been tectonically stable for some 125,000 years! [9] However, when we find such reliance on sea-level modelling we should be cautious, for this is adding a new wing to a castle already built on sand. No-one really knows how big the Pleistocene ice-caps were; no-one knows for certain how much sea-ice melted, or why the ice sheets melted; or why they formed in the first place. It's all cumulative academic opinion served-up as proven fact.

Nevertheless, the specialist view remains that the central Mediterranean has been stable throughout the Holocene. No sunken Atlantis here and no source of mega-quakes either.

However, geologists thought the same about Cascadia – another supposedly stable subduction zone – until evidence of the 1700 tsunami was found along the Canadian coast. The central Mediterranean too, could be a potential site of subduction mega-quakes – but the specialists offer no evidence that they occurred in our timeframe.

The largest documented earthquake and tsunami in this region was the magnitude-7 quake at Taormina, Sicily in 1908, identified as slip along a normal fault rather than subduction; for a detailed account see:

https://www.nature.com/articles/s41598-019-42915-2.pdf
https://core.ac.uk/download/pdf/37834907.pdf

However, we must again note that despite its severe local effects, the Taormina quake did not result in sinking or liquefaction on neighbouring islands, such as Malta and Gozo, nor on Sicily; a more powerful event is needed to match the description given by Diodorus.

Greece and the Aegean Sea

According to Plato, it was during the era of Atlantis that the ground subsided away from around the Acropolis, leaving the rocky outcrop that we know today. Archaeologists find little early archaeology on that site, other than Neolithic pottery around the Klepsydra spring; but a few shards of pottery do not a city make! Perhaps, one day, we may find the remains of an ancient city; and the soldiers, who *"sank into the earth"* at the same time as Atlantis was swallowed by the sea. The Egyptian priest states that Athens was founded before Saïs (an apocryphal thousand years earlier according to the *Timaeus*). Plato's description is informative here: Greece had not yet become *"the bones of the wasted body"* as it is described in the *Critias*. We may ask if there is any archaeological or other scientific evidence to confirm the date of this geological and climate transition.

Greece and the Aegean micro-plate are such a seismically active zone that no-one can doubt that earthquakes and tsunamis have been a regular occurrence as far back as we wish to look. However, to attach a date or magnitude to any specific quake is much more difficult; to link any one of them to a legendary catastrophe is even more problematic. There are so many Greek legends about floods that it would be fruitless to pursue them all here (see Gallanopoulos above). Plato's own description mentions three ancient floods that washed away the soil from around the Acropolis.

An interesting summary-analysis of the Acropolis mound by geologist Callan Bentley may be found at:

https://blogs.agu.org/mountainbeltway/2015/01/17/geology-acropolis/

He remarks that: *"The hill used to be larger, but is being nibbled away over time from the sides. It's an erosional remnant of a much larger thrust sheet"*, but precise dating of these collapses is something that we must await.

Plato's narratives do offer us other clues. The most useful dating indicator is his description of the former temperate climate of Greece. Plato describes: *"the happy temperament of the seasons"* [Jowett translation]. Again, in *Critias* we find: *"the soil benefited from an annual rainfall which did not run to waste off the bare earth as it does today"* [Desmond Lee translation]. This description corresponds best with what we know of the mid-Holocene climate of Europe, which climatologists coincidentally term the Atlantic pollen zone. This name has nothing to do with Plato, but is due to its oceanic characteristics of warm summers and cool winters, without extremes of temperature. For Europe, the transition to the more extreme Sub-Boreal period is placed loosely around 3000 BC. We may note again the synchronism with the emergence of the Nile delta and the establishment of Neit's temple at Saïs, concurrent with the First Dynasty of Egypt.

Pollen evidence reveals the initial stages of agriculture on the northern Thessaly plain from the early Neolithic, extending eventually to the entire plain. [10] A further dating indicator is that we also know there were people disturbing the natural flora. Plato describes hill-tops covered with trees that were cut down in ancient times to roof buildings that were still standing in his own day. In cores taken from the former Lake Viviis in eastern Thessaly the history of the vegetational sequences has been determined. The specialists detect the earliest signs of Neolithic agriculture, which decreased during the early Bronze Age and recovered from the Late Bronze Age onwards. Another recent thesis suggests a similar sequence on the Peloponnese, signifying a wet-dry transition between 5300 BP and 4700 BP [11] Together with the changes in pollen signature, this climate transition confirms the late fourth millennium BC as the era that best fits Plato's description.

We may find numerous indicators worldwide of climate and sea-level change converging around this date. The mid-Holocene corresponds to the period when the Sahara region turned from a grassy savannah to desert conditions. Although some investigators try hard to turn this wet-to-dry transition into a gradual process determined by the Earth's orbit ('orbital forcing') others are clear that the transition was more rapid.

The onset of desert conditions in the Sahara has been determined by cores from the sea-bed off Mauretania. [12] Here it may be shown that the annual deposition of wind-blown desert sand was interrupted between the close of the Ice Age and the mid-Holocene. This is termed the African Humid Period when the Sahara region was vegetated and could sustain large lakes. The study finds that marine sediments lack the temporal resolution to precisely date the onset of desert conditions to within a century; they pin it down to four centuries around 5490±190 BP. One may note (as with many such studies) that the co-authors only dare speak in terms of *gradual* causes; concluding that it occurred when summer insolation crossed a threshold of 4.2% greater than present, thus reducing the effect of the African monsoons.

We may note from the account of the Libyan Amazons' conquests as given by Diodorus that he tells of their defeated enemies taking refuge in a forested region. This offers a clue that we are looking at events from before the dry desert climate set-in and before the displaced Libyan tribes sought refuge in the Nile valley and delta. This would be contemporary with the temples of Malta and the civilisation of the megalith builders along the Atlantic coast. We may have a glimpse here of predynastic Egypt and its Libyan neighbours. This era, we are told, was contemporary with the Atlantian cities in North Africa.

We should perhaps also note the negative dating indicator regarding Thera-Akrotiri. Plato does not mention a volcano. Although the Thera eruption was one of the largest of historical times and despite the tsunami that it triggered, no firm evidence can be cited here that it caused major earthquakes or subsidence on mainland Greece. Even if such were to be found, the pollen evidence noted above suggests that the modern agricultural conditions on the peninsula were already in place long before this Bronze Age eruption.

Conclusions

When we follow strictly the text of *Timaeus* and *Critias* and look to *cross-disciplinary* evidence then the conventional rationalization fails us; it only works if you deem the story to be Plato's own fiction. Scholars of Greek and archaeologists who argue in favour of Thera/Akrotiri have to selectively ignore the absence of climate changes on the Greek mainland at the same era, as well as turning a blind-eye to the synchronism with the foundation of Neit's temple. Both of these indicators suggest a date much earlier: in the mid-Holocene: the Atlantic – Sub-Boreal transition, the mid-Neolithic, or whichever label you prefer to assign it.

The Thera-Santorini-Akrotiri solution for Atlantis that is used as a safe non-controversial explanation simply does not fit the internal evidence of the narratives and comes no closer than some of the ideas promoted by popular enthusiasts; its only saving grace are the academic credentials of those who adhere to it, who might claim to be somehow more qualified or knowledgeable than the rest of us. Not only does the Thera explanation ignore the indicators of date but neither does it fit the geographical description. A city may indeed have been destroyed at Akrotiri, or on Crete, but where is the huge island-continent described by Plato; where is the vast rectangular plain and the canals; and how does it tally with a location beyond the Pillars of Heracles?

It must also be noted that the strongest documented modern quakes, such as the 1755 Lisbon event, did not trigger mega-quakes along neighbouring faults, or sea-level changes within the wider Mediterranean. Still less can they explain how an earthquake or volcanic event could trigger a permanent climate transition in North Africa and Europe. The inference must be that whatever happened around the mid-Holocene was an *exceptional event* of a much greater scale than a magnitude-9 earthquake. Remember, it is the ancient source itself that links the Atlantis story with a catastrophe from the heavens – it is not some speculative creation of modern authors.

This author's recommendation therefore stands. Whenever you find a conflict between the opinion of a modern expert and that given in an ancient text then you should always prefer the source closest to the events. Trust in your most ancient historical source: Plato – or should that be: Sonchis of Saïs?

Notes:

1) The long reigns of the Egyptian gods are found along with the chronicle of Manetho that forms the basis of modern Egyptian chronology; and which gives a total for reigns of gods, demigods, and spirits of the dead covering 24,925 years prior to the dynasties of kings. It is likely that the priests of Saïs followed a similar sacred chronology and therefore included part of this within the Atlantis date.

2) Diodorus Siculus, Library of History, Book III. 52-55. Diodorus is careful not to confuse the Libyan Amazons with the Amazons of Asia Minor who flourished later, a generation before the Trojan War. Be careful! For, we don't know with certainty the era of the Trojan War either. It is another instance of cumulative academic opinion that has morphed into fact.

3) Diodorus Siculus, Library of History, Book IV. 18. 1-7. Diodorus tries to align the Greek and Egyptian legends about Heracles and his pillars but creates only more confusion for us. He says that there was both an earlier and a later Heracles, whose achievements have become amalgamated over time. Diodorus says (in Book III. 7.55) that Heracles entirely destroyed the Libyan Amazons on his way west to the straits to set-up his pillars; and further relates that Heracles in doing so built out into the sea to narrow the straits.

References:

1) National Geophysical Data Center / World Data Service (NGDC/WDS): Significant Earthquake Database. National Geophysical Data Center, NOAA. doi:10.7289/V5TD9V7K
https://data.noaa.gov//metaview/page?xml=NOAA/NESDIS/NGDC/MGG/Hazards/iso/xml/G012153.xml&view=getDataView&header=none

2) https://www.worlddata.info/africa/algeria/earthquakes.php

3) Ambraseys, N. N. The Seismicity of Tunis. Annals of Geophysics, [S.l.], v. 15, n. 2-3, p. 233-244, Nov. 1962. ISSN 2037-416X. Available at:
<https://www.annalsofgeophysics.eu/index.php/annals/article/view/5431

4) Benouar, Djillali. (1994). Materials for the investigation of The Seismicity Of Algeria and Adjacent Regions during the twentieth century. Annals of Geophysics. XXXVII. 10.4401/ag-4466.
https://www.researchgate.net/publication/50300644_Materials_for_the_investigation_of_The_Seismicity_Of_Algeria_And_Adjacent_Regions_during_the_twentieth_century

5) Koster, Benjamin & Reicherter, Klaus. (2014). Sedimentological and geophysical properties of a ca. 4000 year old tsunami deposit in southern Spain. Sedimentary Geology. 314. 1-16. 10.1016/j.sedgeo.2014.09.006.

https://www.researchgate.net/publication/266677764_Sedimentological_and_geophysical_properties_of_a_ca_4000_year_old_tsunami_deposit_in_southern_Spain

6) Garcia, E. et al (2010) Holocene earthquake record offshore Portugal (SW Iberia): testing turbidite paleoseismology in a slow-convergence margin. E. Garcia et al, Quaternary Science Reviews 29, 1156–1172
https://www.academia.edu/2336662/Holocene_earthquake_record_offshore_Portugal_SW_Iberia_testing_turbidite_paleoseismology_in_a_slow-convergence_margin

7) http://atlantipedia.ie/samples/tag/anton-mifsud/
http://atlantipedia.ie/samples/mifsud-dr-anton/

8) Furlani, Stefano & Antonioli, Fabrizio & Biolchi, Sara & Gambin, Timmy & Gauci, Ritienne & Lo Presti, Valeria & Anzidei, Marco & Devoto, Stefano & Palombo, Maria Rita & Sulli, Attilio. (2012). Holocene sea level change in Malta. Quaternary International. 288. 146-157. 10.1016/j.quaint.2012.02.038.
https://www.researchgate.net/publication/235606249_Holocene_sea_level_change_in_Malta

9) Lambeck, K. & Antonioli, Fabrizio & Vulcanologia, Istituto & CNT, Sezione & Roma, & Italia, & Ferranti, Luigi & Leoni, Gabriele & Scicchitano, Giovanni & Silenzi, S. (2011). Sea level change along the Italian Coast during the Holocene projections for the future. Quaternary International. 10.1016/j.quaint.2010.04.026.
https://www.researchgate.net/publication/48329976_Sea_level_change_along_the_Italian_Coast_during_the_Holocene_projections_for_the_future

10) Bottema, Sytze, (1979) Pollen Analytical investigations in Thessaly, Greece; Paleohistoria 21,
https://ugp.rug.nl/Palaeohistoria/article/view/24996/22455

11) Andwinge, Maria, Masters Thesis (2014) Reading Pollen records at Peloponnese, Greece, Stockholm University.
https://www.semanticscholar.org/paper/Reading-Pollen-Records-at-Peloponnese%2C-Greece-Andwinge/7c7dce1e57026e8816f36021d6cb7fa989175050

12) Garcia, E. et al (2010) Holocene earthquake record offshore Portugal (SW Iberia): testing turbidite paleoseismology in a slow-convergence margin, Quaternary Science Reviews, 29 (2010) 1156–1172

https://www.academia.edu/2336662/Holocene_earthquake_record_off shore_Portugal_SW_Iberia_testing_turbidite_paleoseismology_in_a_s low-convergence_margin

This article was originally published as an interactive web-page in 2019 at:

https://www.third-millennium.co.uk/catastrophes-from-atlantis-to-the-a

16   The Stonehenge Builders came from Turkey
   – so what's new?

Upon the release in *Nature* April 2019 of the DNA research by Brace et al: *"Population Replacement in Early Neolithic Britain"* we were treated to sensational headlines in the popular media and online. One headline boldly states: *"Stonehenge builders came from as far as modern-day Turkey, DNA suggests"* [ITV]. Another declared: *"Stonehenge was built by descendants of Neolithic migrants, DNA study shows"* [NBC]. Although the authors of the report were quick to distance themselves from any such spectacular claims, these headlines soon become fixed in the readers' minds. Our first reaction should really be: "so what's new?"

The DNA research reveals that Stonehenge and the other Late-Neolithic stone circles were not built by the first Neolithic farmers to inhabit Britain, but by their descendants, whose DNA *'matched the genetic "signatures" of early farmers in Europe; a group of people shown in previous studies to have migrated from the Aegean coast via Spain and the Mediterranean coast.'*. It now appears that these immigrants did not mix with the older generation of farmers who had arrived earlier in the Neolithic, around 4000 bc.

Up until the 1970s, archaeologists proposed this same origin for the Stonehenge builders based on the artefacts; but in those days cross-dating had placed such influences as recently as 2000

BC. We were confidently told by the archaeologists of that era that the influences of Mycenaean Greece were the inspiration for such grand architecture, which could not possibly have originated among the primitive Brits. These diffusion theories all fell apart when tree-ring calibrated C14 dates pushed this culture back to 3000 BC and earlier. Similarly we have always known that the Indo-European languages, which include Greek and Celtic, as well as the ancient Hittite language of Anatolia, had all originated in the Caucasus and Black Sea region, from which there was an Indo-European language 'dispersal'. All of this may be found in our older text books. So does this new DNA research just confirm what we already knew?

Less well known, and certainly neglected by the scientists, is that this picture also concurs with the narrative that we find in our indigenous myths and legends. The Irish *Book of Invasions* narrates that Ireland was populated by successive waves of 'Greeks': Partholon, Firbolgs and Nemedians who arrived both via Spain and via the Danube route across the North Sea. It seems probable that any such migrants must also have passed through Brittany, Cornwall and South Wales as they progressed.

From Britain we have conflicting stories of origin, mainly surviving in Welsh literature. The most well-known is the myth of Brutus and his 'Trojans' that we find in Nennius. There has always been an unexplained anomaly between the versions of origin given by Classical historians and those in the Welsh 'triads'. All the invasion legends infer that Britain and Ireland were unoccupied and ripe for colonisation.

We find in the commentaries of Caesar and Strabo (both probably citing Poseidonius) a story that the interior of Britain was populated by tribes who claimed to be indigenous to the island. By contrast, the Welsh triads describe the Cymry as originating from the regions around the Bosphorus, whence they came, "after the Flood". The Roman historian Tacitus [in Agricola XI] even remarks on the clear similarities between the people of South Wales and those of Spain, *"...all lead one to believe that Spaniards crossed in ancient times and occupied that part of the country"*, he says.

The Classical historians also recognized that the Celtic gods were the same as those of the Mediterranean pantheon, in all but name; and the Greek poets told us of the close relationship between the Delian Greeks and the Hyperboreans who lived beyond the Celts. The Greek myths tell us of Heracles who, among other achievements, placed his 'pillars' at the Strait of Gibraltar on his way west. In Roman times, the respected Greek author Plutarch is quite clear in his belief that Greeks had colonised the outlying islands of Britain as they migrated there "in the train of Heracles". Can we any longer afford to just dismiss these ancient stories as fiction when they follow the DNA trail so closely?

There is, in the so-called *third-series* of the Welsh Triads, a tradition that Hu Gadarn first brought the Cymry to Britain from the regions around the Bosphorus; but first we are told that they settled in Llydaw, which implies Brittany and the Biscay coast of Gaul. Two other races, the Lloegrwys and the Britons, of the same stock and language, were later allowed to settle. These triads, in the *Myvyrian Archaiology of Wales* have long been condemned as the forgeries of its nineteenth-century author – simply because the twelfth-century manuscript from which he claimed to have taken them – is no longer extant. However, the correspondence with the story that DNA evidence now reveals is quite striking!

The Triad of the three pillars of the Race of the Island of Britain begins as follows:

The first, Hu Gadarn, who first brought the race of the Cymry into the Island of Britain; and they came from the land of Háv called Defrobani [where Constantinople stands] and they passed over Mór Tawch (the German Ocean) to the Island of Britain, and to Llydaw (Brittany) where they remained. The second Prydain...

In the Triad of the three benevolent tribes we are given the same origin for the Cymry; and of the others we find:

*...**The second** were the race of the Lloegrwys (i.e. the dwellers on the river Loire) who came from the land of Gwas-gwn, and were sprung from the primitive stock of the Cymry.*

***The third** were the Britons. They came from the land of Llydaw, and were also sprung from the ancestral line of the Cymry... and the three were of one language and one speech.*

These translations are taken from the *Celtic Researches* of Edward Davies written in 1804. He was a Welsh classical scholar, who could freely make comparisons between the Welsh and Classical languages; so the saying is true: there's nothing new under the sun!

In so far as they may be considered authentic, the Welsh Triads derive mainly from medieval sources in South Wales. Other recent DNA studies have shown that the populations of North Wales and South Wales are quite distant from each other. The modern boundaries of Wales, with its three ancient divisions, are just that mountainous part of Britain that the Anglo-Saxons failed to conquer. The indigenous British inhabitants may therefore be revealed to us as the tribes of North Wales and of the regions further north in England and Scotland.

So perhaps it is overdue that archaeologists and other branches of academic enquiry began to show more respect for the precious stories preserved by our ancestors about their own origins, instead of rubbishing them as 'myths' and legends. If the DNA evidence offers us anything that is new then it should be to provide a semi-historical framework for the events and characters that we find in the oral history left by our ancestors.

Some useful hyperlinks:
https://www.nbcnews.com/mach/science/stonehenge-was-built-descendants-neolithic-migrants-dna-study-shows-ncna995221
https://www.bbc.co.uk/news/science-environment-47938188
https://www.nature.com/articles/s41559-019-0871-9
https://en.wikisource.org/wiki/Triads_of_Britain

British and Irish DNA Provinces

A summary-map showing the most significant DNA provinces of Britain and Ireland based on *Nature Scientific Reports volume 7, Article number: 17199 (2017)*. The full study identifies some 30 regional clusters and considerable overlap. The clusters may be seen to conform to the known older tribal groupings and provinces, apart from southern and central England where population is more homogeneous with continental Europe.

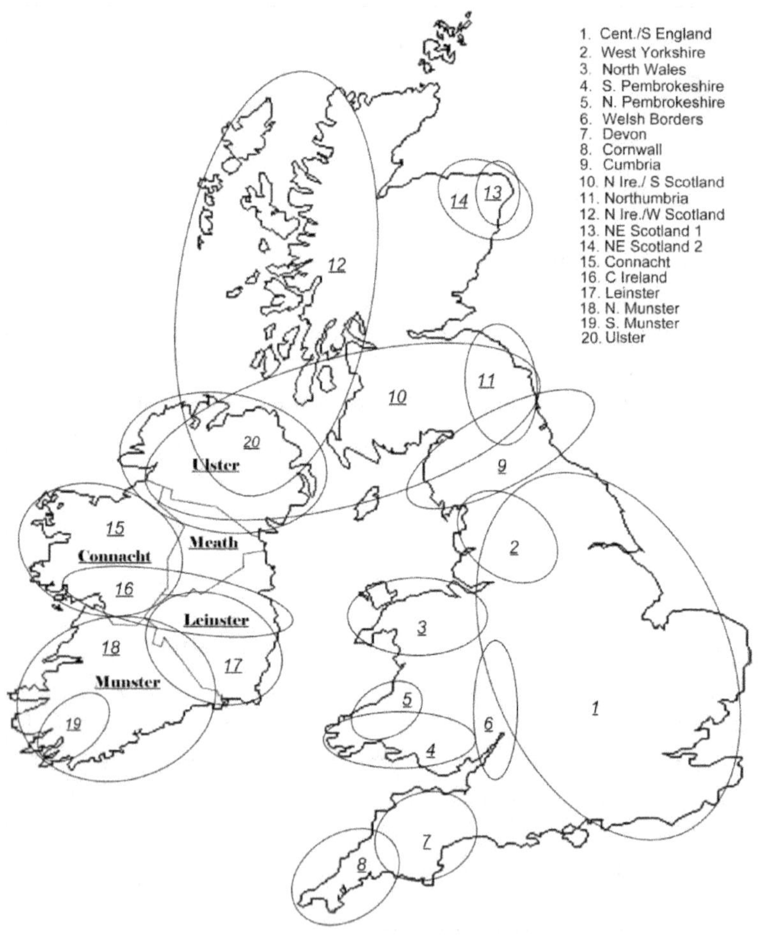

This summary diagram is figure 7.2 from: *Towers of Atlantis*

This article was originally published in 2019 as an interactive web-page on the author's website at;

www.third-millennium.co.uk/stohehenge-builders-dna

17   Who were the Ancient Irish Cannibals?

Perhaps that should be the 'British and Irish Cannibals', but such title does not flow so easily. It begins with the Roman author Strabo (7 BC), who in his geographies compares the man-eaters who dwell on the island of *Ierne* with the similar customs of the 'Scythians' living on the Baltic shores. [1] Diodorus Siculus (50 BC) had earlier described these same people living on the borders of the Scythians, saying:

...and some of these, we are told, eat human beings, even as the Britons do who dwell on the island of Iris... [2]

Now this confusion in describing Britons and Irish, we may perhaps excuse of an ancient author. However, the unique similarity in the customs between Ireland and the Baltic region was, as we shall see below, very real.

Other classical historians also describe these Scythian 'Anthropophagi' as a people who lived on the Baltic coast beyond the Germans. Earlier, Herodotus (450 BC) had described the 'Androphagi' as a distinct race subject to the Scythians and the only tribe in that region to eat human flesh – but then he confounds this by also describing a cannibal practice among another people called the *Issedones!* [3] One must conclude that his Androphagi and Issedones were in fact the same people

– but his two descriptions originate from different sources. Pliny (AD 40) also knew of 'Anthropophagi' and 'Essedones' in what we would now call the Baltic States and Russia – but at that era this entire vast region was simply known as 'Scythia'. [4]

From the fragmentary descriptions given by the ancient writers, the 'man-eating' practice seems to have been more of a ritual than an act of savagery. Solinus (AD 250) gives us more information on the cannibal practices of these Anthropophagi, associating it with a burial rite; as is here described in the somewhat amusing Elizabethan-English translation made by Arthur Golding in 1587. [5]

...It is the manner of the Essedons to follow the corpes of theyr parents singing and...to teare the carcasses a sunder with their teeth, and dressing them with other flesh of beastes to make a feast with them. The skulls of them they bind about with golde, and use them as mazers to drinke in...

Pomponius Mela (AD 40) gives us a very similar account of this ritual, but whether flesh-eating occurred in other more barbarous contexts is not attested. [6] Pliny adds the additional detail of them drinking out of the skulls and also using the scalp hair as a napkin! However, he does not associate this with funeral customs. [7]

Both Strabo and Diodorus considered the customs of the Irish-Britons and 'Scythians' to be equivalent, Strabo indeed saying of the Irish: *"they count it an honourable thing when their fathers die, to devour them..."* [8] It seems unlikely that such unique funerary customs could evolve independently to be so similar in two such distant regions. Strabo also describes the Irish-Britons as following communal social customs involving the sharing of wives between family members. [9]

Other ancient writers also describe tribes (plural) in the interior of Britain who claimed to be aboriginal to the island and who similarly followed this custom of polygamous marriage. [10] The best known of these is the summary in Caesar's commentaries; and Diodorus Siculus also tells us of autochthonous tribes in the interior of Britain who preserved ancient ways. [11]

Move forward now to the late Roman Empire and its battles with Picts and Scots who tried to break into the British province during the reign of Theodosius. Here we encounter another tribal name for the first time. Ammianus Marcellinus, probably referring to events in AD 368, describes Picts, *Attacotti* and Scots as devastating Britain unchecked:

...the Picts divided into two tribes, called Dicalydones and Verturiones, as well as Attacotti, a warlike race of men, and the Scots, were ranging widely... [12]

We see above that these Attacotti are not regarded as Picts from the mainland; and are not Scots either! The Scotti at this era must be regarded as a wholly Irish race. Celtic linguists assess the name: 'Attacotti' to mean: 'very old ones' or aborigines. [13] We do not hear of the Attacotti in the earlier Roman wars with Caledonians and Picts. Perhaps they are hiding under some other name in Ptolemy's list; or they may have lived beyond the Mounth along the west coast and islands, undisturbed by the Romans as they campaigned up the east coast.

Some very unsavoury practices of this same tribe are also mentioned by St Jerome, who describes Atticoti serving in the Roman army of Gaul as: *"a British tribe who eat human flesh"*. [14] He goes on to describe their free social customs: *"like Scots and Atticoti... (they) allow their brides to be promiscuous and live...in communes..."* [15] Now these free social customs are comparable again to those described by Julius Caesar and the other writers as characteristic of aboriginal people in the interior; and this probably refers to more than one isolated tribe in remoter parts of Britain and Ireland (see: Stonehenge Builders DNA).

The assemblage of evidence suggests that we are here looking at an indigenous tribe living in western Scotland and the islands, who had preserved their ancient ways. They would later be absorbed into the kingdoms of the invading Picts, then by the Scots and Vikings; and the ancient customs must surely have been eradicated on Christian conversion. Solinus writing around AD 250 (with additions by later interpolators) predates the earliest mention of the name Attacotti; but he also attributes free social relations to the inhabitants of the Hebrides [*Ebudes*] and Thule [*here probably Lewis*] but somewhat frustratingly he does not mention any form of cannibalism! However, as we have seen above Solinus does describe the practice among the Scythian Anthropophagi, which other writers assure us was the same as that of the 'Britons' living in Ireland! Yes, if you are not thoroughly confused by now then you have not understood the situation! Unfortunately we do not have the ultimate source that these later historians are citing; probably it goes back to the lost Celtic History of Poseidonius of Rhodes, who may have visited Britain in the early first century BC. [16] The table below may help to illustrate the correspondences.

	Strabo	Diodorus Siculus	Herodotus	Pomponius Mela	Julius Caesar	Pliny	Solinus	Saint Jerome
Cannibalism	*	*	*	*		*	*	*
Parent-eating	*		*	*			*	
Skull-drinking				*		*	*	
Polygamy	*				*		*	*
Irish/Scotti	*	*						*
Britons		*			*		*	*
Scythia	*	*	*	*		*	*	
Aboriginal		*			*			

The Picts own traditions would also bring them over from Baltic Scythia as invaders at some time during the first millennium BC; and say that they took Irish wives. [17] They must have encountered indigenous people who (by definition) had been there since the Mesolithic and whose ancestors built the stone tombs and other monuments in Orkney and the Hebrides. In the timeless narrative of the Irish *Book of Invasions*, we are told of the earliest Irish colonists and their battles with 'Fomorii', who are described as *"men out of Scythia and the western isles"*; another coincidence?

Now that DNA evidence has shattered the long-standing dogma about an Iron Age Celtic invasion of Britain we are no longer constrained to find Celts behind every ancient bush! The Picts and the Attacotti were not in any sense 'Celts'; that term should now be used with more caution to describe a linguistic grouping – unless we wish to talk about the Celtic Church or later cultural influences from Ireland.

We find another name in some later Welsh sources: *Gwydll Ffichti* ('Irish Picts'). [18] On Ptolemy's map (c.120 AD) we find in western Scotland: 'Creones', which would be pronounced similar to *Cruithne*: the Irish name for the Picts. [19] These Creones may give us another glimpse of the Attacotti: who as invading Picts living in Ireland had converted to Gaelic language and later brought it into northern Britain.

Now that we are no longer obligated to regard the Pictish language as p-Celtic then the question is open again for scrutiny. In *Picts and Ancient Britons* I made the case that the Picts came from Baltic Scythia, which was after all their own tradition, ignored for so long by Celtic scholars; and that they bought with them a Finnic language. It may be that contact between northern Britain and the Baltic goes back a very long way, even as far back as the Mesolithic when the route was open via the North Sea land bridge. This is an inevitable consequence if we regard the Attacotti as an aboriginal group – that is what aboriginal implies.

To find the origins of the cannibalism and the free social customs of ancient Britain and Ireland then we must look further back into what we know of the Neolithic. DNA evidence now tells us that there was an influx of invaders from the southern Steppes into southern Britain (the 'beaker people') during the Late Neolithic. This population replaced the earlier farmers and their culture during the Stonehenge era of the third millennium BC. (see: Stonehenge Builders' DNA) We must regard the arrival of the northern Picts as a much later incursion around the era of the broch-builders. As for the aboriginal Britons, it is clear that much of their way of life survived in isolated pockets of the older culture. We should perhaps regard the population replacement as analogous to the late Roman Empire when barbarians from the east broke in, not so much to destroy the culture as to become a part of it; but eventually they overwhelmed the older population.

An interesting facet of this Late Neolithic incursion is the revolution in burial monument style. These changed from the long-barrows and court tombs towards high-status burials in single barrows. Perhaps we should examine what we know of these Neolithic burial customs from the archaeology.

All around Britain and Ireland are found the dolmens of an older culture dating from the Early and Middle Neolithic. In Scotland and the north of Ireland we find the Court Cairns: stone-covered chambers with a crescentic forecourt. Further south in Britain we find instead the dolmens or 'cromlechs' similar to those of Brittany and Iberia. These were built of larger megalithic stones, formerly covered by a mound but lacking the forecourt; and in southern Britain we find the earthen long mounds such as Wayland's Smithy and West Kennet. All these structures likely performed a similar function as community ritual sites (churches or crypts?) rather than as tombs.

Intact skeletons are virtually unknown from the early Neolithic period and preservation does not seem to have been important. Excavators report evidence of the scattering of human as well as animal bones in the area around the cairns; as if unwanted older bones were routinely disposed-of to make space for the new arrivals. A helpful summary of the changing burial styles may be found at:

http://www.spoilheap.co.uk/burial.htm

Archaeologists also report that the skulls are 'under-represented' among the disarticulated remains found at these sites. To give perhaps a typical example from southern England one may consider the nineteenth century excavation of the West Kennet long barrow in Wiltshire, which showed that it too once possessed a crescentic forecourt typical of the Scottish and Irish styles. When the mound was first excavated in 1859 the jumbled bones of around 46 individuals were found. A more detailed description of the excavation may be found at:

http://www.sacred-destinations.com/england/west-kennet-long-barrow

We must set aside our natural disgust at the notion of cannibalism and consider that to these descendents of early North-Europeans it was a normal part of their nurture. It may be that the ritual devouring of the deceased parent (or perhaps just of the father) and the retention of selected bones as relics, was a form of ancestor reverence. By consuming some small piece of the body the parent literally became a part of the child; and it seems that the adornment of the skull as a drinking vessel was also part of a ritual rather than pure barbarism.

We may visualise that in the forecourt of the ancient dolmen, the cannibalistic ceremony took place as it is described, with the body ritually torn and bitten by the assembled extended family; and the meats of parent and other beasts being ritually mixed and cooked over a fire. The body itself was not cremated, but left in the open dolmen-crypt. After a prescribed interval, the bones, particularly the skull, might be retrieved and retained by the family as a relic of the revered ancestor who would therefore always be with them. We see this for example in the West Kennet long barrow, which was only sealed in the Late Neolithic when the people of the earlier Neolithic had retreated to the western margins of Britain. We must await more evidence from archaeology as to exactly what happened to the aboriginal people of southern Britain, be it natural catastrophe, disease or genocide; perhaps a succession of all these things. The descendents of these earliest people, we now know from DNA, are still with us today; in the west of Scotland, North Wales and Yorkshire; perhaps in the north of Ireland too.

It is only by putting together a cross-disciplinary picture that we may perceive this view of the ancient society – one cannot find it solely from archaeology or the accounts of any individual classical writer. By putting together the overlapping historical references to ritual cannibalism and communal social structure, with the DNA and the archaeology, then we may perhaps experience an 'aha' moment. So that's what they were doing!

References

1. Strabo Geography IV, 5, 4
2. Diodorus Siculus, Library of History, V, 31, 2-3
3. Herodotus, Histories, III and IV, various references
4. Pliny, Natural History, IV, xii, 88
5. Solinus, Collecteanea Rerum Memorabilium, additamenta, 15, 13 and 22 (11-17)
6. Pomponius Mela, De Chorographia, II, 8-10
7. Pliny, Natural History, IV, xii, 81
8. Strabo, Geography, IV, 5, 4
9. ibid
10. Julius Caesar, The Gallic War, V, 12 and V, 14
11. Diodorus Siculus, Library of History, V, 21, 5
12. Ammianus Marcellinus, Library of History XXVII, 7, 8 & 8, 5
13. Rivett, A.L.F. and Smith, C., The Place Names of Roman Britain, Batsford, London (1979)
14. St Jerome, Against Jovinian, 2, 7
15. St Jerome, Epistola, 69
16. Inferred from a remark by Strabo in: Geography, 2, 4, 2
17. Principally Bede, Ecclesiastical History, 1, 1 — but others have the same story.
18. Triads of Arthur, 36
19. Ptolemy, Geography, C, Liii

Longer extracts of all the literary sources are given in:
Picts and Ancient Britons.

The author's other Pictish Studies are not published in this volume and may be found via the sub-menu at:

https://www.third-millennium.co.uk/features

18   Lyonesse – Lost!

The legend of a lost land called Lyonesse, submerged off the rugged Cornish coast between Land's End and the Isles of Scilly has found its way into medieval and popular folklore based on a small number of primary sources. In addition, we find related legends of a sunken city lying off the Brittany coast, together with the comparable Welsh and Irish legends of lost lands. Usually commentators on the theme of Lyonesse are diverted into a discussion of Arthurian literature, which then leads to a trivialisation of the legends as fiction and myth. However, in the 1950s the Welsh geologist Frederick John North thoroughly examined all the medieval sources for the various myths of sunken cities. His conclusion was that they were inspired by a memory of a single ancient event that subsequently became dispersed to various locations. This article will therefore concentrate on an exploration of the earliest root sources.

The version of the Lyonesse legend that is most often quoted comes from the fifteenth-century itinerary of Cornwall by William of Worcester (1415-1482) who tells us of:

...woods and fields and one-hundred-and-forty parochial churches [parishes] *all now submerged, between the Mount and the Isles of Scilly.* [1]

The 'mount' here refers to the tidal island of St Michael's Mount off Marazion, around which submerged forest deposits are found. Its Cornish name means 'the grey rock in the wood', implying that it was formerly dry land surrounded by forest.

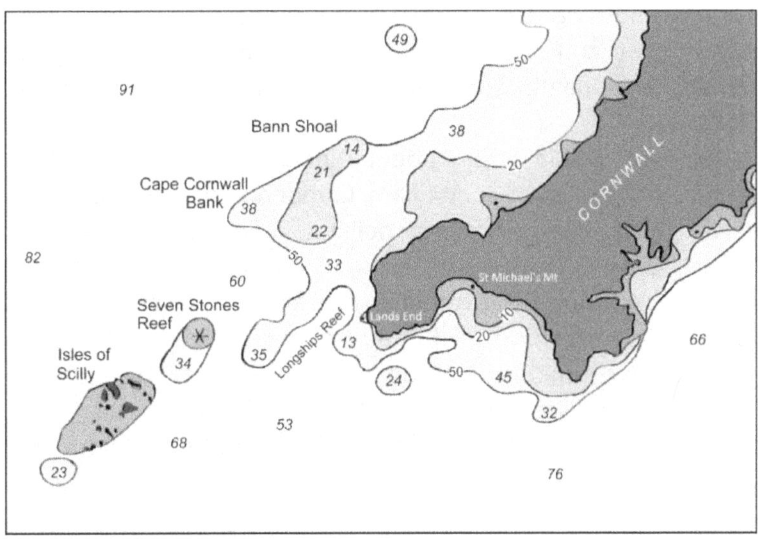

The bathymetry of Lyonesse summarised to show only the 10, 20 and 50m contours, based on UK Admiralty Charts 1123 and 2675. The sea-bed to the south and west of Land's End drops away sharply, but it may be seen that flat areas exist between 20-50m depths to the north of Penwith; and also a shallow rise between the Seven Stones and Longships reefs. However, the Isles of Scilly are separated from the mainland by a deeper channel.

From the Elizabethan era we have the version of the map-maker John Norden (1548-1625) who in his *Description of Cornwall* refers-to: *'the supposed drowned lande of Lioness'*. [2] These lands, he says, according to tradition and the stories told by seafarers, were 'swallowed-up by the devouring sea'. Norden's version would localise the lost land to the vicinity of the Seven Stones reef off Land's End, which in the Cornish language was called Lethowsow: 'milky-white rocks'.

The monk Florence of Worcester also describes a coastal inundation in his chronicle of events for the year 1099. [3]

On the third of the Nones of November the sea overflowed the shore, destroying towns and drowning many persons and innumerable oxen and sheep.

There is nothing to directly connect the 1099 event with Lyonesse. It could record a severe winter storm surge, or perhaps a tsunami, like that from the Lisbon earthquake of 1755.

Lyonesse would attain popular notoriety due to the brief description published in William Camden's *Britannia* of 1586, which most commentators believe influenced map-maker Richard Carew's *Survey of Cornwall*. [4][5] Camden refers to '*Lionesse*' as formerly extending out from the rocks of Land's End, where once stood a lighthouse to guide ancient mariners.

Yet other timeless traditions would tell of the survivor *Trevilian* who escaped the sudden inundation on his horse, ahead of the waves. There must, of course, always be a survivor to witness the tragic events! Some versions however, would associate him with the 1099 event recorded by Florence; all these details become rather mixed-up in the popular retelling of a very few real facts.

From Brittany, A similar legend is told about the lost city of *Ker-Is*, which in Breton means 'the low-city'. The tale describes a city submerged in the Bay of Douarnenez, as told by the local fishermen. There are many variants, which would attest to how old the story is. The earliest extant version is recorded in Albert le Grand's '*Lives of the Saints of Armorican Brittany*' published in 1637. [6]

The story remembers King Gralon, who built a city and palace on land reclaimed behind sea-walls and sluices. His wayward daughter Dahut stole the keys and in her folly she opened the sluices allowing the sea to pour in. Gralon was warned by an appearance of Saint Gwenole and tried to rescue his daughter, but she fell from the horse (or was thrown) into the sea. Le Grand's version is somewhat simpler; the saint warns Gralon that God intends to punish the inhabitants for their sins and he flees on horseback before a great storm inundates the city — but the blame still falls upon the reckless princess! The death of Gwenole is elsewhere recoded as AD 532 and would therefore serve to date the associated legends to the sixth century.

For those who would focus upon the Arthurian links there is a parallel that Douarnenez lies in the Brittany province of Finistère, which is of course the French equivalent of English: 'land's end'. It is therefore postulated that perhaps the memory of a lost land was transferred from Brittany to Cornwall by scholarly monks during the early evolution of the story. [7]

The precise etymology of the name *Lyonesse* is obscure; the legend itself is likely older than its name. It is found in the fifteenth-century fiction of Sir Thomas Malory. In *Le Morte d'Arthur*, Lyonesse becomes the birthplace of the character Tristan; this in turn builds-upon timeless legends that he came from the lands around Saint-Pol-de-Léon, on the north coast of Finisterre. This fiction augmented older folklore in circulation about King Arthur and brought him into a contemporary medieval setting.

While we are less concerned here with King Arthur and his knights, he cannot be entirely divorced from the enquiry. The historical Arthur (or Ambrosius Aurelianus) is believed to have flourished in the sixth-century dark-age following the withdrawal of Roman rule from Britain; so much so that the entire interlude has come to be known as the 'Arthurian Period'. We find semi-legendary accounts of this era in the pseudo-history of Nennius, but evidence for the existence of a real 'Arthur-figure', who resisted the Saxon invasion, is elusive. [8] His existence, at this period, relies upon the dating of the Battle of Badon Hill (near Bath) as described in the contemporary account of Gildas and dated at AD 516 by the Welsh Annals. [9] On such fragile connections then, rests the supposed submergence of Lyonesse at this same era.

The interchange of folklore and myth between Cornwall and Brittany does have solid historical foundations. We know from Caesar's commentaries that the maritime Gaulish tribes, the Veneti and Osismi, were regularly trading with Britain, probably as part of the ancient tin trade. [10] Later sources tell us of the migration of Britons from Cornwall and further north, into Brittany, from the late Roman period onward into the fifth and sixth centuries, to escape the invading Saxons. [11]

So we see a chicken-and-egg problem. Were the legends of sunken cities and lost lands transferred from Cornwall to Brittany along with the Cornish settlers? Or could the legends actually be older and reached Britain through earlier exchanges with the maritime Gaulish tribes? A third possibility is that they evolved independently from common ancient roots.

The association of Lyonesse, with the Isles of Scilly and the submerged sea-bed between the islands are an additional factor to add complexity. These were most likely the Cassiterides or Tin Islands of classical writers. The antiquarian William Borlase in 1753 noticed lines of stones running between the islets around the Isle of Samson and believed that they could be man-made field walls; indeed it is still possible to walk between some of the islets at the very lowest tides. [12] Scilly has Neolithic chambered cairns similar to those of Penwith and so it reasonable to conclude that they were built by the same people; and by inference that the 'field walls' may be of similar age. A Neolithic passage grave at Er-Lannic on the Brittany coast now lies on the beach; and another lies half-submerged off Westward-Ho!, Devon; this must attest to a rise of sea level by at least a few metres since these monuments were built. However, this is not of the same order as a regression by over 60m that would be required to link the Isles of Scilly to Cornwall.

A recent archaeology survey entitled *The Lyonesse Project* studied the sea bed around the Isles of Scilly. [13] Part of their conclusions was that the present configuration of the islands was *'largely formed by the end of the early Bronze Age'* but that the intertidal zone was formerly more extensive. A submerged forest was discovered in 2006 by a diving expedition in St Mary's Roads, between the main islands; the deposits showed signs of human clearance of the oak/birch forests. Radiocarbon dates from the same project showed them to date from the 'later Mesolithic' between 5310-5050 cal BC. The study did not seek former shorelines beyond 10m below modern sea level, although these must exist from earlier in the Holocene if current theories of Ice Age and Holocene sea-level change are accepted. During the last glacial maximum, sea-levels were some 120m lower than today.

Although the authors of the Lyonesse Project annexe the popular name and summarise the legends (to display thoroughness) the study shows the typical trivialisation of the folklore. It is distressing to see wording such as: *"Although much of this can be dismissed as fantasy..."* within a scientific report. Folklore is not fantasy; it is degraded and half-forgotten history and as such deserves more respect than it receives in academic circles.

In 1957, at the end of a long career as keeper of geology for the National Museum of Wales, Frederick John North examined all the legends of sunken lands around the British coast and published them in his landmark study *Sunken Cities*. [14] His thorough analysis would conclude that Lyonesse was one of several localisations of a single ancient inundation that occurred somewhere along the west coast of Britain. The degree of variation and repetition between the local legends would suggest a long period of oral preservation before being recorded; and this confusion would then be compounded by misunderstandings in medieval map-making.

So is it possible to add anything new on the subject of Lyonesse? Anyone who wishes to pursue the medieval folklore, or the influence that it had on Arthurian literature, will find many other discussions and articles. However, here, I shall explore a much older inspiration for all these legends based on a cross-disciplinary approach.

An interesting intrusion that is not usually associated with Lyonesse comes from the sixth century historian Procopius in his *History of the Wars*. Writing of the Gothic Wars of AD 536-532 he offers us a remarkable aside into the myths of the people living along the Brittany coast at that era. [15] Although he lived in far-away Palestine and was confused in his geography of Britain he is one of our few historical sources for this dark period. He refers to an island called *Brittia*, which he presumed to lie beyond the former Roman province of Britannia, inhabited by both Britons and Angles. This latter detail is enough to tell us that 'Brittia' was probably the region of Southern Scotland between the two Roman walls: British Strathclyde and Anglian Bernicia; from here too, he tells us, the Britons were migrating to Gaul.

Procopius describes a custom among the fisherman and sea-traders along the coast of Gaul, who at that era were subject to the Franks. Since ancient times they were charged with the duty to ferry the souls of the recently deceased to the west of Britain. We may be sure that such a voyage was purely symbolic – for the crossing took less than an hour when transporting the souls, yet a full day and a night in normal circumstances. The souls of the dead, he says, *'were always conveyed to this place'*. We may recognize here the Celtic 'otherworld', the Tír Na nÓg or Tir fo Thuinn of Irish myth: the land beneath the wave.

The significance here is that Procopius wrote at the same era as the historical Arthur is thought to have lived. He only includes the myth as an aside to his history because *'it was constantly being published by countless persons'* and he did not wish to gain a reputation for ignorance of the custom. He makes no mention of King Arthur, or Lyonesse, or of any recently submerged cities. We may see that the legends of lost lands beneath the western sea were already so ancient by the sixth century that they had penetrated the religious mythology and funeral rites.

Other strange references from Celtic sources supplement this mythology of a sunken land. In the strange *'Prophecies of Merlin'* as included in Geoffrey of Monmouth's *'History of the Kings of Britain'* we find another reference to a time of lowered sea levels. [16] The prophecies predict that the strait which separates Britain from Gaul will be reduced to a narrow channel, such that a man on either shore might shout to one on the other. If we accept Geoffrey's claim to have taken his history from a lost Welsh book then it must recall an ancient epoch, when the sea level in the English Channel was much lower. According to Merlin's final prediction, winds and storms would prevail when the sea returns. We find these same fears voiced in the thirteenth-century: *Lament for Llewellyn Ap Gruffudd*, which compared the death of the last independent king of Wales to a natural calamity: '*...why, O my God does the sea not cover the land?*' [17] We may ask what lies at the root of such strange ideas. Note that in the Celtic legends, the flood is an overflow of the sea; this is not the Biblical Deluge of Noah.

Other legends of sunken cities are abundant around the Welsh coast. As with Ker-Is these are too many and various to explore in detail when our focus should remain upon Cornwall, but the similarities must be highlighted. Most well known from the Welsh Triads is the legend of Cantrae'r Gwaelod (the Lowland Hundred) in Cardigan Bay, also known as Maes Gwyddno (Gwyddno's Plain); we are told that Gwyddno was the king of Cardigan at that time. The legend recalls sixteen cities, submerged in Cardigan Bay, supposedly during the time of Ambrosius. The blame is laid upon Seithenyn, the warden of the sluices (and an arrant drunkard according to another triad) who forgot to close the gates one night. [18]

The similarities with the Cornish and Breton legends are immediately apparent; the cities lie below sea level behind sea-walls and Seithenyn replaces Dahut as the villain who opened the gates. However, the drowning of sixteen cities is more reminiscent of the version of Lyonesse recorded by William of Worcester, who gives a hundred and forty submerged 'parishes'. In both cases we may see that the memory is of a *substantial* tract of *flat* land that was *rapidly* inundated. The preservation of precise numbers in a legend also confers a degree of authenticity; a fiction need only mention 'many cities' or refer to a vague area of submerged land.

The Welsh triads and the prose stories of the Mabinogi, so the Welsh linguists tell us, began to be written down shortly after the departure of Roman authority from Britain. Another triad, the Exeter Triad, is preserved only in Latin translation within the thirteenth-century *Chronicle of Wales*. [19]

These are the kingdoms which the sea destroyed. The kingdom of Teithi Hen...That kingdom was called at the time the Realm of Teithi Hen it was between St Davids and Ireland. No one escaped from it, neither men nor animals, except Teithi Hen alone with his horse.

The second kingdom was that of Helig son of Glannog, it was between Cardigan and Bardsey and as far as St Davids. That land was fertile and level and was called Maes Maichgen; it lay from the mouth (of the Ystwyth?) to Llyn, and up to Aberdovey.

The sea destroyed a third kingdom: the kingdom of Rhedfoe son of Rheged.

We may see again the localisation of the root-legends that should by now be familiar. Like Trevilian in Cornwall and king Gralon in Brittany, here it is Teithi the Old who escapes on a horse. The second kingdom is recognisable as another version of Cantrae'r Gwaelod in mid-Wales, but was ruled by Helig rather than Gwyddno. We are not told where Rhedfoe's kingdom lay; or indeed whether all these inundations were contemporary. However, Rheged was the kingdom of the North Britons in Galloway and Cumbria – the 'Brittia' of Procopius.

More details of the Welsh legends are preserved in the stories of the Mabinogion; medieval prose compositions that are clearly much older in origin. In the mabinogi of *How Culhwch won Olwen*, we again encounter Teithi Hen described as he: *'whose kingdom the sea overran'*; again supposedly in the era of Arthur. In the story of *Branwen Daughter of Llyr* we hear of the giant Bran and his army, who was able to walk from Wales to Ireland by crossing only two rivers, *'but thereafter the sea widened and overflowed the kingdoms'*. [20] We see again a vague memory of a much wider submergence event than the loss of a single city. The confused details within the Mabinogi betray a long period of oral transmission and reworking by storytellers, who lacked an understanding of the history upon which they were based.

Although F.J. North could convincingly explain some localisations of the submergence legends by mistakes in medieval mapping, he concluded that the degree of divergence between the stories showed the era of the sunken cities to be much older. The Exeter Triad, albeit a Latin translation of a lost Welsh source, must be far older than any medieval map. We may see that it was not just one city but entire 'kingdoms' that were lost to the sea. The triads must be pre-Christian and even pre-Roman in inspiration; they were intended for oral transmission and only frozen in their present forms when eventually written down in medieval Wales. Many more historical triads as well as prose must have been lost over the centuries.

Another source of information is the Roman geographer Strabo (63 BC – AD 25); he cast doubt upon the comments of Poseidonius who wrote a world history in the early first century BC. Poseidonius certainly visited Britain and Gaul himself, but mere fragments of his work survive in citations by later authors. In discussing the rising and settling processes of the Earth, Strabo remarks that Poseidonius believed Plato's account of the sinking of Atlantis and the description of Britain given by Pytheas. [21] The ultra-conservative Strabo would accept neither! He also attributes the migration of the Germanic Cimbri (around 110 BC) to *'an inundation of the sea that came on all of a sudden'*. [22] It is unfortunate that we don't possess more from the histories of Poseidonius; we may wonder if he too knew about the Celtic legends of lands and cities, lost to the sea.

At 335 BC we are told of Alexander the Great and his meeting with Celtic emissaries on the Danube, prior to embarking on his Persian wars. [23] He enquired of the Celts what they feared most, and received the surprising reply that they feared no-one, only that the sky might fall on their heads! This has been likened to an early version of the Irish Oath, which voices similar fears and continues: *"If we fulfil not our engagement…may the sea, overflowing its borders, drown us"*. These same fears recur in the first-century epic: *Táin Bó Cúailnge* (The Cattle Raid of Cooley). This would suggest that the dread of a sudden inundation by the sea must be at least as old as the era when Celtic-speakers reached Ireland, certainly long before the time of Alexander.

Elsewhere in Irish legend we find references to *'the illimitable seaburst'* and Ireland's *'three great floods'* together with their associated folklore. These perhaps formed a triad, similar to those from Wales. [24] The legends tell of Clidna's Flood (*Tonn Cliodna* or 'Cleena's Wave') together with Ladru's Flood and Baille's Flood. Other than the drowning of the unfortunate princess we have little additional detail. However, unlike the Welsh triads we are told that the floods occurred at different times. It should be apparent that any extreme waves that afflicted the Irish coast must also have struck along the west of Britain.

The case for an Iron Age invasion of Britain and Ireland by continental Celts is no longer as certain as it formerly seemed. Linguists now prefer to describe only 'Insular Celtic' and 'Continental Celtic' languages, rather than a racial grouping; so the memory of an ancient inundation must be older than the era when the Celtic languages began to separately evolve. Modern DNA evidence now places the influx of people with Steppe-ancestry into the core-Celtic regions as early as 2800 BC, with further genetic evidence that they did not intermingle with the earlier Neolithic population until several centuries later. [25] Historical comparisons suggest that native cultures persist until there is a language shift and people begin to forget their former identity (or perhaps it is suppressed by the dominant culture). It is reasonable to suppose that the legends of 'sunken cities' were preserved among the native population of the Atlantic coast rather than by Celtic-speaking intruders from the east. The linguistic evolution therefore pushes the origin of the flood legends back into the Neolithic and early Bronze Age, before the Celtic languages diverged.

The 'Celtic' mythology, of which Lyonesse is a part, survives from the oral history of the Bards. From Poseidonius, via Strabo and Diodorus, we have a description of the Bards as one of the three elites in Gaulish society, along with the Druids and the Vates. The Bards were poets and singers who recorded history in their oral chants. [26] From Caesar's commentaries we learn that the Druid order originated in Britain and that their initiates were forbidden to write anything down. [27] This, in-part explains the incomplete preservation of the flood stories, which have survived Roman and Christian suppression of the old religion. We should be grateful that anything remains, but their historical sequence has been lost. We may also see that Lyonesse and the Welsh sunken cities are associated with the era of the hero Arthur, but we cannot know for certain when that was. The Arthur whom we encounter in the Welsh Mabinogi bears little resemblance to the Arthur of later literature; it may be that he is a composite-figure of legend, to whom numerous events, including the submergences, have become attached.

All around the coasts of Britain and Ireland, sea-level researchers find the physical evidence of submerged forests. Although these date from various eras, we may see that they are all of Neolithic age or earlier. Young trees and other land vegetation cannot grow in a salt-water environment. The preservation of tree-stumps, fallen trunks and peat layers beneath beach-sand imply that they grew on dry land and the mature forests were rapidly covered by sand, before they could decay. Furthermore, we can be sure that much of the modern coastline has changed little since Roman times. We know from Diodorus Siculus (first-century BC) that St Michael's Mount was already a tidal island called Ictis, to which ships would arrive to buy the Cornish tin. [28] Radiocarbon dates for the submerged forest around St Michael's Mount give ages between 4000 and 6000 BP, all far older than the sixth-century when the submergence of Lyonesse is usually placed. [29]

Submerged forests from Cardigan Bay give radiocarbon dates ranging from 5500 BP for the tree stumps around Borth, to 3500 BP at nearby Ynyslas. In the Bristol Channel the deposits at Stolford date from 5398-5020 BP and those around the dolmen at Westward-Ho in Devon gave 6500 BP. Other dates from the submerged forest deposits of the east coast are typically older, associated with the flooding of the North Sea land-bridge. By the very nature of submerged forests deposits we can infer that these were rapid transgressions rather than a gradual encroachment of the sea. The date of submergence cannot be older than the forests. The evidence from the western coasts therefore suggests a cluster of dates in the Middle Neolithic around five thousand years ago, together with a later episode around 1500 BC.

Could folk-memory really survive from such remote eras? If we look further back, to the Mesolithic, we find the acceptance by archaeologists of the *Maglemosian Culture* since the early twentieth century; a community of hunter-gatherers who lived both in northern England and Denmark, before the flooding of the North Sea separated them sometime between 9,000 and 6,000 years ago. However it is only quite recently that archaeologists have dared to consider that during the period of 3,000 years when Doggerland was above the waves, people may actually have lived on what is now the floor of the North Sea! [30] This is rather obvious when you think about it. It is only a small leap of imagination from there to see the possibility of settlements of similar age on the now submerged land to the west of Britain. It is a question of when, not if. It is a much greater leap of imagination for archaeologists to perceive a fairy-tale Arthurian Lyonesse in the flints and pottery that they find. Some extreme theorists such as Koudriavtsev and author David Furlong would even seek to locate Plato's Atlantis deep in the Celtic Sea off Cornwall at the close of the Ice Age. [31][32]

Perhaps there is no need to go back quite that far to find the roots of the Lyonesse legend.

As with the Lyonesse Project mentioned above, we may see that archaeologists and others are quick to 'dismiss' legends and traditions when they don't fit their conclusions, yet they will employ them where they confirm their views. Take this example, from a paper of 2010, on the subject of Irish monuments and myths. [33]

Societies frequently explain the origin of distinctive features in the landscape by weaving them into a mythological narrative...Some mythic places exist primarily in the imagination. England's Camelot and the Irish Tír na nÓg are as much utopian states of being as physical locations. Where imaginary worlds feature in the lore of a society, their interface with day-to-day life is often linked with mysterious features in the local countryside.

This quotation is from an archaeologist, who is actively seeking to equate legend with archaeology, yet even here we see the 'dismissive' mind-set; he still prefers to see ancient fiction rather than degraded and jumbled memories of real places. One may find statements of this kind in many specialist papers. It is sometimes difficult to distinguish whether this predisposition is the true opinion of an academic researcher, or if it is merely the safe thing to say to ensure that their papers are published. Until this attitude changes we are unlikely to see any real attempt by archaeologists to investigate the submerged archaeology of the offshore basins.

Conclusions

We may see that the legends of submergence fall into two groups: the first type describes the overwhelming of large tracts of land with towns and cities; the second is a lesser event, typically with more detail, remembering the inundation of a single low-lying city. The Lyonesse legends would therefore fall into the first grouping.

We see in the Lyonesse story, an example of a composite myth, formed by the merger of similar-sounding ancient events and characters. Somewhere around the coast of Britain there was a rapid flooding of a low-lying coastal city (or perhaps more than one). However, the details and characters have been combined, during the evolution of the folklore, with memories of a much earlier and wider submergence of land off the western coasts. It is this earlier event that lies at the root of the legends of sunken kingdoms and the land of the dead beneath the waves. It may even be that details and characters associated with more recent tsunamis or severe winter storms have also become entangled in the retelling of the ancient stories. Such is the nature of oral myths and legends. It is only when a story is finally written down that it becomes fixed.

Table of Correspondences	Lyonesse	Ker-Is	Cantrae'r Gwaelod	Teithi Hen	Rhedfoe
Kingdoms with many cities	*		*	*	*
Single city below sea level		*	?		
Survivor on a horse	*	*		*	
Sea walls and sluices		*	*		

The physical evidence tells us that we should seek the lost lands in an ancient era. Perhaps as far back as the end of the Ice Age; but *more likely* the submergence of the kingdoms and settlements, was linked with the same mid-Neolithic event that left the submerged forests around British and Irish coasts. We may see that the submergence, whatever the cause, was *sudden and catastrophic* rather than a gradual transgression from polar ice melt; neither could it describe a tsunami, as the submergence was permanent. To find the flat plains with cities, as they are described, then we would have to look further-out on the sea floor, perhaps as far as 30-50m depths. Perhaps one or more cities closer to the modern coastline survived this flood and were protected behind sea walls, only to fall victim to a later inundation.

It may be that there is no lost city lying off Cornwall and it should be sought in one of the other places; but as Frederick North suggested, the memory of a lost land has been transferred to Land's End in the retelling. If we should look on the navigation charts then we may see that there is no flat region on the sea-bed off Land's End that could hold a hundred and forty settlements; more likely these could lie off the north coast of Cornwall, or remember a wider submergence around the west of Britain. Rather than drowned Christian churches from the sixth century we may find Neolithic sites of worship, like the dolmens and passage graves of Penwith and Scilly.

I would like to think, that if Frederick North had also had the benefit of radiocarbon dates, back in the 1950's, then he too would have come to similar conclusions. It is absurd to suggest that ancient people would invent a fiction about sunken cities, merely to explain the presence of tree trunks on a beach, or some rocks lying off the Cornish coast. Even more unlikely that people in diverse regions should each independently adopt a belief that the sea might suddenly rise-up and drown the land. After all, no-one would dream of such a thing happening today, would they?

Notes and References

1 Harvey, John, ed. (1969) *Itineraries [of] William Worcester*: edited from the unique MS Corpus Christi College, Cambridge, Oxford Medieval Texts. Oxford: Clarendon Press. ISBN 0198222033

2 Norden, John (1650) *A Topographical and Historical Description of Cornwall*; Published in facsimile by Frank Graham, Newcastle Upon Tyne, 1966

3 *The Chronicle of Florence of Worcester*, translated by Thomas Forester, A.M. Bohn, 1854.

4 Camden, W (1586) *Britannia*, published in online translation at: https://www.visionofbritain.org.uk/travellers/Camden/8

5 Carew, R (1602) *The Survey of Cornwall*; republished in 2000 by Tamar Books, Redruth, Cornwall

6 Le Grande, Albert (1637) *Lives of the Saints of Armorial Brittany* [link to text] https://gallica.bnf.fr/ark:/12148/bpt6k5038760

7 The Isles of Scilly, Lost Peaks of Lyonesse? In *Seidkona's Hearth*: http://www.pollyanna-jones.co.uk/?m=201401

8 Nennius, 56

9 Gildas, 25, 3

10 Caesar, III, 7-11

11 Procopius, History of the Wars, VIII, xx, 7-9

12 William Borlase, *An Account of the Great Alterations which the Islands of Scilly have undergone since the Time of the Ancients*, Philosophical Transactions of the Royal Society, No. 48. 1753

13 Dan Charman, Charlie Johns, Kevin Camidge, Peter Marshall, Steve F Mills, Jacqui Mulville, Helen M Roberts (2014) *The Lyonesse Project: a study of the coastal and marine environment of the Isles of Scilly (OASIS ID cornwall2-58903)* [data-set]. York: Archaeology Data Service [distributor] https://doi.org/10.5284/1025045

14 F.J. North (1957) *Sunken Cities*, University of Wales Press, Cardiff

15 Procopius, History of the Wars, VIII, 7-9; and 47-58

16 Thorpe, Lewis (1966) *Geoffrey of Monmouth – The History of the Kings of Britain*, Penguin, Harmondsworth

17 Conran, Tony (translator) (1986) *Welsh Verse*, Poetry Wales Press, Bridgend, p 163

18 Bromwich, Rachel (1950) Cantrae'r Gwaelod and Ker-Is, in Fox & Dickens (eds) *The Early Cultures of North West Europe*, pp 217-241

19 Bromwich, Rachel (1978) *Troiedd Ynys Prydein – The Welsh Triads*, University of Wales Press, Cardiff

20 Ganz, Jeffrey (1976) *The Mabinogion*, Penguin, Harmondsworth

21 Strabo, Geography, 2.3.6

22 ibid

23 Arrian, I, iv

24 Peate Cross, T. & Harris Slover, C. (eds) (1935) *Ancient Irish Tales*, George, G. Harrap & Co Ltd, London.

25 Furtwängler, A., Rohrlach, A.B., Lamnidis, T.C. et al. Ancient genomes reveal social and genetic structure of Late Neolithic Switzerland. *Nat Commun* 11, 1915 (2020). https://doi.org/10.1038/s41467-020-15560-x

26 Strabo, Geography, 4.4.4;

27 Julius Caesar, The Conquest of Gaul, VI, 16, 5, 13-14

28 Diodorus Siculus, Library of History, V, 22, 2-4

29 Sources of these radiocarbon dates are listed at: https://www.third-millennium.co.uk/submerged-forests-britain-ireland

30 https://humanities.exeter.ac.uk/archaeology/research/projects/title_892 82_en.html

31 http://www.subtleenergies.com/ormus/wg/atlan4_e.htm

32 Furlong, David (1997) *The Keys to the Temple*, Piatkus, London

33 O´Sullivan, Muiris, Megalithic tombs and storied landscapes in Neolithic Ireland, pp53-66 in Martin Furholt et al (eds) *Megaliths and Identities: Early Monuments and Neolithic Societies from the Atlantic to the Baltic*, 3rd European Megalithic Studies Group Meeting 13th – 15th of May 2010 at Kiel University

This article was originally published as an interactive web-page in 2020 at:

https://www.third-millennium.co.uk/lyonesse-lost

19   Irish God Kings and a Nine-Month Day!

Following the article in Nature by Cassidy et al (2020) there has been renewed interest in the Irish mythology surrounding the construction of Newgrange, and its legendary builder Echu Ollathir, also known as Dagda: "the Good God". The interest stems from DNA evidence of in-breeding within members of the Irish royal family, an incidence of which is actually described within the legends. So, if we are to validate one part of a legend by modern science then we should ask: on what grounds do we dismiss all the other statements within it? Here is an analysis of some of the astronomy within those legends.

Genealogists sampled 44 genomes from remains discovered within the deep recesses of the Newgrange passage grave; one of which they identified as the adult son of an incestuous first degree (brother-sister) conception, typical of those within Egyptian royal families. [1] Remains from other nearby passage graves were also found to be close relatives of this same individual. This raises the possibility of a dynasty of elite 'god-kings' among the early Irish, analogous to those of Egypt. The royal remains showed a DNA profile typical of the Western Neolithic population of Atlantic Europe (light-skin and brown eyes) that expanded westward, along with agriculture, from Anatolia and the Black Sea region around 6-7,000 years ago. [2] However, other burials also showed evidence of Irish hunter gatherer DNA (darker skin and blue eyes) – indicative that some of the population had survived from the earlier Mesolithic inhabitants. [3]

In the metrical *Dindshenchas* (place-name legends) we find origin-stories for the tombs at the Bend in the Boyne; in particular the building of the Newgrange and <u>Dowth</u> passage graves by the legendary kings of the Tuatha Dé Danann.

Newgrange prior to the modern reconstructions (Macallister 1931)

The unreconstructed mound of Dowth

The Dagda and his son Aengus óg (Macc Oc) were god-kings of the Tuatha Dé Danann, colonists from 'Greece' according to the *Book of Invasions*, who settled Ireland via Iberia early in prehistory after Ireland had become deserted. They fought many battles with earlier colonists. Aengus was the son of Dagda and the river goddess Bóann (after whom the river Boyne takes its name). In essence, the story recounts that Dagda commanded the sun *to stand still for nine months*, so that the child might be conceived and born within a single day. From this derives his name Aengus óg or 'Aengus the Young'. As he grew-up, Aengus learned that he would inherit nothing from his real father but he was determined to possess Newgrange. So he sought permission to live in the mound for a day and a night. The following day Aengus declared that a day and a night was equivalent to all days and all nights. By this trick he became the owner of Newgrange.

The building of nearby Dowth is attributed to another king of Ireland named Bressal. There came a great disease or 'murrain', which killed both men and animals, leaving only seven cows and a bull. Bressal attributed the plague to the will of the gods and gathered the people together at Bruig Na Boínde (Newgrange). He set them a task to build a great tower in a single day, so that he might reach the heavens and end the plague. The men agreed to work for a day, but this was a trick. Bressal's sister was a powerful druidess and she cast a spell to hold the sun high in the sky creating an *endless day* for them to build the mound. The spell was broken only because the king committed incest with his sister. Day became night and the men refused to work any more. They returned home leaving behind Dowth: the place of Darkness. '*Dubad* [darkness] *shall be the name of this place for ever*', declared the druidess.

I am unsure whom I quote, but it is certainly a truism that archaeologists and other academics will cite myths and legends whenever they support their conclusions; yet will dismiss or ignore them as just 'myths' when they do not. So we may wonder whether many of the other legends surrounding Newgrange might have some validity that could be verified by science – especially those that describe the strange astronomy.

The era when the Boyne valley monuments were built is now firmly established by radiocarbon dates since the excavations of the 1970's. Although usually termed 'passage graves' it cannot be proven that they were constructed solely as tombs. The Newgrange excavations, which resulted in the reconstructed monument, revealed a precise alignment of the main passage and 'roof-box' towards the midwinter solstice sunrise. Organic material between the stones gave a date of 3150±100 BC. [4] According to the authors of the 2020 DNA analysis this coincided with the era when Ireland was rapidly colonised, displacing an earlier native population.

The primary source for the legends surrounding the building of Newgrange and the other Boyne Valley monuments are the *Metrical Dindshenchas*, from the twelfth century *Book of Leinster*. The Dindshenchas, served as a memory aid for the bards to recall the facts of more detailed stories; these were then learned as poetry and recited orally. We may suppose that many more details of the ancient stories have been lost.

The episode is described in the Dindshencha of *Boand II*, verses 8-9 as here translated by Edward Gwynn. [5]

*Thither came by chance the Dagda
into the house of famous Elcmaire:
he fell to importuning the woman:
he brought her to the birth in a single day.
It was then they made the sun stand still
to the end of nine months — strange the tale —
warming the noble fine grass
in the roof of the perfect firmament.*

We must trust to the skill of the translator and of the bards who preserved these stories over the millennia. The other report in *Dubad*, describing the building of the Dowth mound, is rather less precise; recalling simply, an endless day.

His sister came to him, and told him that she would stay the sun's course in the vault of heaven, so that they might have an endless day to accomplish their task. The maiden went apart to work her magic. Bressal followed her and had union with her: so that place is called Ferta Cuile from the incest that was committed there...

Tales of gods and spells may serve to satisfy the general population as to the motives of their rulers. However we may see here a degraded memory of the construction of the passage graves by astronomer-priests who clearly understood real astronomy. In the surviving versions of the Dindshenchas we also detect Christian influence; the building of a mound to reach the heavens is likened to Nimrud and the Biblical Tower of Babel. Clearly, the later Irish bards who recorded the poems for us had no idea what their ancestors might be describing.

Another Irish story from the *Yellow Book of Lecan* is *The Wooing of Etain*, which also takes place in the time of the Tuatha Dé Danann, a generation after the building of Newgrange. It elaborates on many of the themes found in the Dindshenchas. [6] Here again we are told of the Dagda:

...for he it was who performed miracles and saw to the weather and the harvest, and that is why he was called the Good God.

Of the Dagda's relationship with Bóann, the wife of Elcmar we are told:

The Dagdae charged Elcmar with great commissions, so that nine months passed like a single day, for Elcmar had said he would return before nightfall. The Dagdae slept with Elcmar's wife, then, and she bore him a son who was named Oengus...

Dagda took his son to be fostered in the house of Mider at Brí Leíth, west of Newgrange; Mider descended from the earlier invaders of Ireland, the Fir Bolg. So Aengus grew-up believing Mider to be his father, but when he was old enough Mider told him of his true parentage. Aengus asked Mider to take him to Dagda, to be acknowledged as his son and to grant him the land that was his due. Dagda was willing, but said he could not grant Bruig Na Boínde to Aengus, as it belonged to Elcmar and he did not wish to anger him further. Dagda himself then suggested a ruse: that Aengus should occupy the mound for a day and a night and lay claim to it. So Aengus threatened Elcmar with his life but spared him on condition that he could occupy the mound overnight, to which Elcmar agrees; but Aengus is not to release it to Elcmar unless the matter is first put to the judgement of the Dagda. Aengus argued that Newgrange was his by right in return for having kept his promise to spare Elcmar; ownership for a day and a night was to mean all days and all nights. The Dagda so gave his judgement and placated Elcmar with a grant of good land elsewhere.

The story continues a year later when Mider comes to visit his foster-son. He asks Macc Oc for a reward of the fairest woman in Ireland, Etain daughter of Ailill. However, Ailill will not give the girl to Mider and demands a price of gold and silver from Macc Oc. He also insists that he ask the all-powerful Dagda to clear twelve plains on his land in a single night. Thus, the myth explains how the twelve great rivers of Ireland were set to flow towards the sea, where no rivers had flowed before!

Ailill releases Etain to Mider and he returns with her to Brí Leíth, but there is an obstacle. Mider has a first wife, Fúamnach, a powerful druidess; in her jealousy she turns the beautiful Etain into a scarlet fly that follows Mider wherever he goes. The complex story may be summarised in its most important elements. Fúamnach summons up *seven years* of lashing winds that blow the fly away, but at the end of that period the fly returns and settles on the tunic of Macc Oc at Newgrange. The angry Fúamnach summons seven more years of lashing winds to blow the fly away, this time for ever. Why is the precise period of seven years so important to be stated?

The spirit of Etain is later reborn as the daughter of a warrior named Etar and we are told:

Etain was conceived in the woman's womb and was born as her daughter. One thousand and twelve years from her first begetting by Ailill until her last by Etar.

We see again, the recording of precise numbers in a myth where no such precision is required for a fictional tale. The number is woven into the myth to preserve its era for future generations and gives us a clue that the poem was composed a thousand years after the time of Newgrange, when the astronomy upon which the original narrative was based had already become arcane.

The second half of the Etain story need not greatly concern us; it describes how she is reborn as a young girl and how Etar is tempted by an otherworldly visitor, to visit *Mag Már,* 'the great plain': one of the many names of the Celtic Otherworld. Thus we see a typical theme found in many Welsh and Irish myths, where a well known hero-figure is woven into a semi-fictional tale to preserve the knowledge of the Land of the Dead.

If this strange astronomy were found only in the Irish legends then it might be easier to dismiss them as just 'myths' – but this is not the case; there are parallel supporting references from other cultures.

In Search of Thule

The Roman writer Pliny describes the phenomenon of the Arctic winter and he leaves us a further clue: that the phenomenon of the midnight sun could sometimes occur in the island of Mona (Anglesey); an island that was sacred to the British Druids. [7] Anglesey lies on a similar latitude to the Boyne, just across the Irish Sea. The ultimate source for this belief could only have been the Druids of Anglesey themselves. Pliny continues that the same phenomenon also occurred in the remote northern Island of Thule. [8]

In his description of Britain, Pliny describes the location of Thule, saying that the crossing to the island began from the largest of the outlying British islands, called *Berrice*:

The most remote of all those recorded is Thule, in which as we have pointed out there are no nights at midsummer when the sun is passing through the sign of the crab, and on the other hand no days at midwinter. Indeed some writers think this is the case for periods of six months at a time without a break. [9]

Notwithstanding the opinion of Gerald of Wales in his *History and Topography of Ireland* that Thule was an island well known in the East, but completely unknown in the West, it is clear that classical geographers from the Mediterranean did not know where Thule actually was. [10] The remark of Pliny above would suggest that it was Iceland, but other geography derived from the voyage of Pytheas would seem to be describing the Outer Hebrides. [11]

By the time that Procopius wrote in the sixth century AD the north was better known. His references to Thule are clearly describing Scandinavia, the northerly extent of which remained unknown. [12] He tells us of the *Scrithiphini* or 'skiing Finns': the Saami people who lived there. In his account he describes the forty days of darkness around midwinter and their fears that the sun might never return:

And when a time amounting to thirty-five days has passed in this long night, certain men are sent to the summits of the mountains – for this is the custom among them – and when they are able from that point barely to see the sun, they bring back word to the people below that within five days the sun will shine upon them.

Unlike the Irish, the northern Saami would have regularly experienced the summer midnight sun and would be unconcerned by it. However, they also experienced the cold winter months of darkness. Why would they be so worried that, in some years, the sun might not return? As the polar seasons are today a regular and predictable calendar phenomenon, we may ask why the Scrithiphini should fear that the sun might not behave as expected. This may be a folk-memory of the same abnormal seasonal phenomenon that is described in the Irish myths of Dubad.

The proto-Finns lived to the south of the Saami; and in the earliest known times were spread across an area around the Gulf of Finland, where today we find northern Russia and the Baltic states. In their *Kalevala* songs, they too preserved an oral tradition about the disappearance of the sun. In Runo XLVII we find the myth of the theft of the sun and moon. The shaman *Louhi* seeks revenge on the people of Kalevala and sends disease to kill the people and their cattle. She hides away the sun and moon. The good shaman *Väinämöinen* heals the sick and forces the evil Louhi to return the sun and the moon to the sky.

As the ancient homeland of Finns and Estonians lies on the same latitude as Britain and Ireland then they should have experienced the same seasonal abnormalities as are described by the Irish. In any rationalisation for such astronomy, this must imply a temporary dip of the zodiac below the horizon, such as occurs in an Arctic winter. Unfortunately the Kalevala myths are quite timeless.

Norse mythology has the myth of Ragnarok; the *Fimbulvetr* is a prophecy of a harsh winter that will last for three years without a summer in-between as the wolf Skoll swallows the sun. Ragnarok will bring with it many other upheavals and human struggles. Again the myths are timeless, but we may assume that the ancestors of the northern Germanic peoples lived in southern Scandinavia and the Baltic coasts; and their expansion north is recorded in historical times.

From as far away as China we find the myth of Kung Kung (Gong Gong) in the time of the legendary empress Nu Wa (Nüwa) in whose reign a cosmic event is recorded. There are many variants of this story.

We are told of a battle between Gong Gong and Zhuan Xu for the imperial throne, which became so violent that the sky tilted-over! Gong Gong collided with one of the pillars of heaven about which the sky rotates. The balance of the world was restored only when empress Nüwa repaired the heavenly pillar, but her patch was imperfect! We are told that the world became tilted towards the southeast while the sun and moon shifted to the northwest; and this supposedly explains why all the rivers of China flow towards the east. Subsequently we hear many legends of how the earliest emperors struggled to control the floods of China's great rivers.

The era of Nüwa is only loosely defined by the Chinese legendary chronology, but some time before 2850 BC gives us an indicator for when these cosmic events supposedly occurred.

The *seven year rhythms* in the climate and its use as a unit of time – as found in the Etain story – are widespread in Welsh and Irish myths. Seven-year rhythms from other parts of the world are most widely understood from the Biblical story of Joseph, with its seven good years followed by seven lean years. Whether the Joseph story recalls real events or whether it be an apocryphal tale to warn future generations what to do, should it ever happen again, is worthy of debate.

The seven year climate rhythm is also found in the Babylonian Flood story in the *Epic of Gilgamesh*, which also dates from the third millennium BC. [13] The god Anu gives a prophecy of a famine so deep that the dead will outnumber the living. He says to Ishtar:

If I do what you desire there will be seven years of drought throughout Uruk when corn will be seedless husks. Have you saved grain enough for the people and grass for the cattle...

An Egyptian record of a seven year famine is also preserved on the Famine Stela at Sehel Island in the River Nile, which tells of a food scarcity in the reign of King Djoser. Egyptologists now prefer to recognise the Biblical story as one among many memories of seven-year famines throughout near-eastern culture. Again, the dating of Djoser is early third Millennium BC, around 2750 BC.

Whenever we encounter memories of similar phenomena among geographically dispersed ancient peoples then we should ask how this can be so? It cannot just be dismissed as 'myth' as that would require diverse ancient peoples to come together to standardise their beliefs just to confuse later generations! Why would they choose a seven-year rhythm that just happens to conform to the known characteristics of the Earth's rotational wobble? Why not choose five or ten years? Why not just say 'many years'? Such explicit numbers are always an indicator that a memory of something real has survived as a fossil within the oral myths.

We may also recall the Biblical references in the *Book of Joshua* to the sun 'standing still' in the sky; [14] a story which would have been well known to the later Christian bards who recorded the Irish legends. They would surely have recognised the old stories as confirmation of their Christian teachings – otherwise the myths would have been suppressed like so many other pagan beliefs.

This theme of the rotation standing still was echoed in the amusing 1930's Alexander Korda film of H.G. Wells': *'The Man Who Could Work Miracles'*. The hero, George Fotheringay is granted god-like powers. [15] In his folly, he orders the world to stop spinning and unleashes chaos – at which point God puts an end to the experiment and restores the normal rotation! Common sense should tell us that this is not a valid mechanism for a physicist to consider. Notwithstanding the vast energy that would be required to halt the diurnal rotation without demolishing the planet, what force is there to then set it spinning again without similar chaos? Think again.

A more plausible mechanism to explain the ancient astronomy would be a 'nodding' of the axis that could temporarily bring polar conditions further south. In the Irish myths, the sun appeared to remain high in the sky (i.e. it did not set); as clear a description of a long arctic summer day as we could wish, but it is somehow extended to the nine months so that Mac Oc could be conceived and born within a single day. How seriously should we take this precise statement?

In my own earlier research for *Atlantis of the West* I remarked on the probable early dating of the Irish legendary-history, which went against the then-prevailing scholarly consensus. This held that the legends in the Irish Book of Invasions were mere literary fictions, supposedly created no earlier than the arrival of Irish-speaking Celts during the Iron Age. If I may quote my own earlier words:

> *...the story of Etain is set unequivocally within the era of the Tuatha De Danann...this fits quite well within the chronology previously established, from the Book of Invasions, suggesting a date of about 2800 BC for their arrival in Ireland. It is almost as if the references to the strong winds were deliberately woven into the story in order to preserve its era.* [16]

However, with the benefit of the archaeological dating it might now be better to condense the era of all the Irish invaders described in the Book of Invasions to the few decades around the building of the passage graves, perhaps as early as 3150 BC and contemporary with the First Dynasty of Egypt. Now that DNA evidence has discredited the older dogma about Iron Age Celts it is easier to see that the insular Irish Celtic language and culture arrived much earlier, perhaps during the early third millennium BC – and that the oral legends were recorded by the bards in an early form of the Gaelic language.

The concept of the long polar day and night was certainly understood by the geographers of Greece and Rome; and we may presume that they gained their knowledge from first-hand experience by natives of more northerly lands. The obvious suspect here is the lost astronomy of the Druids. [17] Pliny describes the long polar days and nights much as we recognise them today, but no-one at that era could have actually experienced conditions at the North Pole.

At the North Pole six months of continuous daylight occur; reducing such that at latitude 66.5°N, at the Arctic Circle, the sun just skirts the horizon on the day of the solstice. Just south of the Arctic Circle is a region of about six degrees in which *Civil Twilight* occurs as the sun dips just below the horizon; plus a further six degrees of *Nautical Twilight*. If such were to be observed as far south as Anglesey or the Boyne valley then it might have been considered as daylight. To find such conditions today one would need to experience the 'white nights' of St Petersburg, or anywhere along a roughly 60°N latitude – Helsinki, Stockholm, Bergen or Shetland. This phenomenon occurs only on the days around midsummer solstice.

Of course in the southern hemisphere the opposite phenomenon of an Antarctic winter would simultaneously be occurring. However, as there was no human presence in those regions until more recent millennia then it would have been witnessed only by penguins! Logically an extended polar night should also have been experienced in the northern hemisphere, either before or after the 'endless-day'. We are not told that *Dubad* was anything other than the return of normal days and nights, but the other northern myths described above would seem to describe a polar night.

Consider how this could happen. We may immediately rule-out here a true change of *obliquity*, that is, a change to the tilt of the Earth's principal axis in space. Notwithstanding the prohibitive energy requirement, an increased obliquity such as would drag the Arctic Circle further south would only result in a six-month polar day being experienced further south – not a nine-month day as is described. For the ancient observer to experience a day that lasted nine-months then it would imply that the axis was somehow precessing around the orbit, prolonging the solstice conditions.

The Free Nutations of the Earth
The Earth has two modes of free wobble should the rotational axis become disturbed. The better known is the *Chandler wobble*, a body-wobble routinely measured by modern geophysicists. This could be excited by internal rebalancing within the Earth. It would be accompanied by sea-level variations, floods and changes to the geodesy of the land. However, it does not seem to be a good model for the events described in the Irish legends – although we do find these phenomena described in other myths, such as the Chinese stories related above.

A better candidate for the Newgrange myths would be the *Free Core Nutation* (FCN); a second theoretical oscillation known to geophysicists but which is vanishingly small on the Earth today. It could only occur should the axis of the outer crust and mantle become displaced from that of the fluid core, but would require an initial external force to excite the motion.

The core nutation is a motion of the instantaneous axis of rotation *in space*, which would be accompanied by only a small body wobble. Thus it would be experienced by humans as an irregular motion of the sky; just as we perceive the daily rotation of the Earth as if the sky were rotating rather than the planet upon which we stand. The theoretical period of the core nutation has now been estimated from the various models and equations at 432 days; [18] and once triggered it should persist for thousands of years. As a free transient motion, its amplitude must decay exponentially to rest, being damped by the fluid friction at the core-mantle boundary. On the Earth today the internal and external axes are almost perfectly aligned and the motion is extinct, leaving us with only equations and theory.

To an observer on the surface during such an episode, the sky would appear to be rotating daily about a point, an excitation pole, which would itself be describing an irregular circuit about the true pole of the sky. This is our modern celestial pole – or such as we could project it back allowing for secular variations. The celestial pole is determined in the instant of the event that disturbed the axis. The excitation pole would be describing a (retrograde) circuit around the celestial pole sometimes enhancing and at other times suppressing the seasons over a cycle of roughly seven-years; gradually spiralling inward as the oscillation is damped. Since the rate of decay is exponential then its amplitude would have reduced most rapidly in the early decades after the disturbance.

It is important to distinguish between the motion of an excitation axis and that of the principal axis. The principal axis could only be changed by an external force from space, but we need not debate here what this force was; the geophysicists will speak vaguely about 'sources of excitation' within their theoretical models. The development since the nineteenth century of the Kelvin-Hough model of the Earth as an elastic shell containing fluid resulted in a better understanding of the theoretical oscillations. The second mode of wobble would require the axis of the shell to be tilted one way, with a compensatory displacement of the fluid core in the opposite sense; the axes of the two layers would then precess about the principal axis in a conical motion, interacting at the core-mantle boundary.

Geophysicists can only study the tiny oscillations of the Earth today. The Chandler wobble is measured only in arc-seconds; the core nutation so small as to be barely detectable! The modern amplitude is only of the order of 20-30cm. The same equations would apply to a much larger motion; however we cannot be certain that the same constants of elasticity and fluidity would apply in extreme circumstances. There is every reason to believe that a significant displacement of the axis of the outer shell might become *chaotic* in nature – completely unpredictable. The events that we see described in the Irish myths would demand an erratic nodding motion of at least 6-8° of latitude, in order for polar twilight conditions to occur as far south as Ireland. Some may find this hard to accept.

However, before you are tempted to raise the speculation flag and declare all the above to be pseudo-science, think again. *No it isn't!* The conditions that must prevail under the regime of a core nutation were first derived by William Hopkins as early as 1839 which later models of the Earth as an elastic shell surrounding a fluid core all cite; if I may quote him here.

> *Since the motion of the interior fluid cannot be subjected to observation, it would be useless to make the substitution of numerical values in equations. We may remark, however, that the motion of the axis of instantaneous rotation of the fluid will be exactly similar to that of the axis of the shell, and of the same order, as is easily seen by comparing the two equations just mentioned.*
>
> *...By this inequality, therefore, alone the points P and P' would describe circles about the same centre in the same periodic time, with radii in the ratio of $Y1: Y2$ and differing in angular position by $180°$. The motion now described is that which would obtain if no extraneous disturbing forces acted on the spheroid, and the axes of instantaneous rotation of the shell and fluid should be separated by a small angle. It is a case of rotatory motion which has not before been investigated.*
>
> *...In addition to the above motions of precession and nutation, the pole of the earth would have a small circular motion...* [19]

Later researchers would censure Hopkins for flawed mathematics and geology, but then negative criticism is what academics do best. Hopkins is a forgotten pioneer. His overall conclusions about the nature of the free core nutation were largely correct. However, he believed that his investigations showed the interior could not be fluid, because it was not observed to actually behave that way. Geophysicists neglected the existence of the core nutation for 130 years until the second half of the twentieth century. When in the 1980s I began to consider its relevance in ancient astronomy and myths, it still didn't have a recognized name.

Conical Chaos!

The attitude of the principal axis in space need have been displaced only a little by the initial excitation. The misaligned axes of core and mantle would then be linked by fluid viscosity at the core-mantle boundary. Geologists have only recently discovered continent-sized regions at the base of the mantle, linked to volcanic hotspots, along which magma could flow freely. [20] However, although the derived constants and equations may hold for the small-amplitude oscillations measurable today, it is possible that for a larger displacement the spatial coning motion could become chaotic and unpredictable.

The concept of chaotic motion within a system of conical motion goes as far back as the experiments of Robert Hooke, which pre-date Newton's physics. As discussed in the paper by *Argentina et al*, Hooke must have observed this chaotic motion in his own experiments to study Kepler's planetary orbits and so restricted his experiments with conical motion to only small angles. [21] Chaotic rotational motion is observed by astronomers for small non-spherical satellites and comets, such as that of Saturn's moon Hyperion. An analysis of how chaos theory might apply to the Earth's wobble must await some practical observations, such that the geophysicists have a real rather than a theoretical oscillation to consider. Otherwise, to quote geophysicist Kurt Lambeck:

> *...[the] parameters are all poorly known, making any discussions of damping mechanisms, and for that matter of the excitation processes of an as yet largely unobserved oscillation, little more than speculation.* [22]

Since the 1980s when Lambeck wrote this, great strides have been made in the study of the tiny motion of the axis on the Earth today. Latest measurements give a period of 431.4 days.

[23] This interval of 432 days and multiples thereof are also found in ancient Indian calendars and Babylonian astronomy; and this hardly seems coincidence. [24] Perhaps the ancient astronomers have left us the real observational data that modern geophysicists need to fully understand the motion.

Not to forget the Swans!

There is one more astronomical clue in the Irish legends and it comes from the various references to swans – yes, *swans*! Irish mythology is replete with stories about swans; the characters often transform into swans, as in the story of Etain.

In *The Children of Lir*, the children are changed into swans by a spell that traps them for 900 years – another precise number. By the end of that time the age of the Tuatha Dé Danann had long passed and a Christian priest changes the swans back into children – but they age rapidly and die.

Another myth, *The Dream of Aengus* again concerns Aengus óg. He dreams of a girl named Caer and falls in love with her; but she and her sisters are transformed into swans every second Samhain (October 31st) and they then remain in the form of swans until the following Samhain. Aengus sets-out to find the swan that is Caer and to undo the spell! So again we see precise numbers in a myth. What might be the significance of this three-year period?

The swans that over-winter in the Boyne valley and in western Scotland are migratory Bewicks swans. Every spring the birds fly to Iceland to experience the polar summer north of the Arctic Circle, where there is a brief period of abundant food for their offspring; but each autumn around mid-October they return. It may be that every three years or so, in the unusual conditions of a nutation, the swans did not experience their usual climate trigger to fly north and believed that they were already in their summer home. Thus their anomalous presence in Ireland over the summer months was remembered in so many myths and legends.

It is easy to see that, for the common people, the swans were seen as somehow bringing with them the abnormal seasonal conditions. This may have been how it was explained to the people by the bards in their songs. As such, the ancient Irish astronomy has come down to us through oral transmission by later generations of bards who could not understand what-on-earth their forebears were talking about!

Conclusions

Firstly, it is necessary to accept that myths are a degraded memory of something real and are not just fiction. So, to attempt a summary of the astronomy that the ancient Irish legends may record, based on the trail of cross-disciplinary clues presented above, would be something like as follows.

Late in the fourth millennium BC an 'astronomical event' disturbed the Earth's rotation. It separated the rotation axes of the crust and mantle from that of the fluid core and it began to wobble. The first mode of wobble would have decayed within around twenty years and relative normality would be restored. This catastrophic phase was completed (but perhaps still within living memory) by the time that the Tuatha Dé Danann colonised a much depleted Ireland.

However, the core nutation would still have been ongoing and causing seasonal climate variations that could not be easily predicted. The Irish and also the Britons set out to understand it by building astronomically aligned monuments to track the sun's motion in the sky. The oscillation must initially have been chaotic, causing large excursions of the excitation pole and resulting in irregular arctic summers and winters as far south as Ireland. This would demand a core nutation of at least 8° of latitude – although the principal axis (the obliquity) need only have changed by a small angle. This is the era that is recalled in the Dindshenchas and other origin myths about Newgrange.

It seems that within a generation the rate of exponential decay was such that the oscillation had settled into a more stable pattern, as predicted by the geophysical theory; and we see the references to the seven-year rhythms in the climate as the nutation-seasons alternately suppressed and enhanced the annual seasons. Exponential decay is a powerful effect, such that the amplitude of the nutation should have decayed to become barely noticeable to the common people within a few hundred years. However, its effect on the sun, moon and stars would still have been observable by competent astronomers until it was fully extinguished, some time in the mid first millennium BC.

It is little use to seek hard evidence of a transient wobble that has long since decayed to rest. What should we expect to find? It should have left its mark in the climate and sea level record; and perhaps in tree-rings, ice-cores or in the record of volcanism and earthquakes. All these clues do exist but are attributed to other causes by the various specialists. The era at which to look for evidence is clearly established by the archaeological dating of Newgrange to the period 100 years either side of 3150 BC.

So much then for cross-disciplinary analysis of ancient fossil myths; where else might you find Irish mythology discussed alongside incest, DNA, radiocarbon dates, geophysics and migratory swans? Will any specialist dare to put names to the human remains found in Newgrange and restore, perhaps, a piece of ancient Irish history that is older than the pyramids?

Relevant Hyperlinks

https://www.nature.com/articles/s41586-020-2378-6
http://www.carrowkeel.com/sites/boyne/macnewgrange.html
http://www.carrowkeel.com/sites/boyne/dowth1.html
https://www.egypttoday.com/Article/4/54056/Famine-Stela-A-piece-of-Pharaonic-diary
https://commons.wikimedia.org/wiki/File:The_Man_Who_Could_Work_Miracles_by_H._G._Wells_(The_Illustrated_London_News,_Summer_1898).pdf
https://ui.adsabs.harvard.edu/abs/2001jsrs.meet..142M/abstract
https://www.sciencealert.com/scientists-have-found-surprise-structures-wrapped-around-the-centre-of-earth
https://www.timeanddate.com/astronomy/civil-twilight.html
https://www.timeanddate.com/astronomy/midnight-sun.html
https://www.transceltic.com/pan-celtic/swan-celtic-mythology
https://www.third-millennium.co.uk/spirals-on-long-meg

References

1 Cassidy, L.M., Maoldúin, R.Ó., Kador, T. *et al.* A dynastic elite in monumental Neolithic society. *Nature* **582,** 384–388 (2020). https://doi.org/10.1038/s41586-020-2378-6

2 Haak, Wolfgang (June 11, 2015). "Massive migration from the steppe was a source for Indo-European languages in Europe". Nature. Nature Research. **522** (7555): 207–211. arXiv:1502.02783. Bibcode:2015Natur.522..207H. doi:10.1038/nature14317. PMC 5048219. PMID 25731166.

3 Mittnik, Alisa (January 30, 2018). "The genetic prehistory of the Baltic Sea region". Nature Communications. Nature Research. **16** (1): 442. Bibcode:2018NatCo...9..442M. doi:10.1038/s41467-018-02825-9. PMC 5789860. PMID 29382937.

4 Ray, T. P. (1989) The winter solstice phenomenon at Newgrange, Ireland, accident or design? Nature, **337,** 343-5

5 Gwynn, Edward (ed.), "The Metrical Dindshenchas", Royal Irish Academy Todd Lecture Series, Hodges, Figgis, & Co., Dublin ; Williams and Norgate, London; Part 3 (1913)

6 Ibid, part 4 (1924)

7 Pliny, Natural History, II, LXXVII

8 Pliny, Natural History, IV, xvi, 104

9 ibid,

10 Gerald of Wales, History and Topography of Ireland, 50

11 Strabo, IV, 5, 5

12 Procopius, *History of the Wars*, Book VI, XV, 4

13 The Epic of Gilgamesh, from the translation by N.K. Sanders, Penguin, Harmondsworth, 1972

14 Joshua, 10, 12-14

15 H.G. Wells 'The Man Who Could Work Miracles', a short story

published in *The Illustrated London News*; Summer 1898.

16 Dunbavin, Paul (2002) *Atlantis of the West*, Constable & Robinson, London, ISBN: 1-84119-716-5 (page 235)

17 Caesar, *The Gallic War*, VI, 14

18 F. Roosbeek, P. Defraigne, M. Feissel, V. Dehant, The free core nutation period stays between 431 and 434 sidereal days, *Geophysical Research Letters*, vol. **26**, no. 1, pages 131-134, January 1, 1999

19 Hopkins, W. (1839) Researches in Physical Geography, *Phil. Trans. R. Soc.* London, **129**, pp 381-423

20 D. Kim, V. Lekić, B. Ménard, D. Baron, M. Taghizadeh-Popp
Sequencing seismograms: A panoptic view of scattering in the core-mantle boundary region
Science 12 Jun 2020: 1223-1228

21 Médéric Argentina, Pierre Coullet, Jean-Marc Gilli, Marc Monticelli and Germain Rousseaux, 'Chaos in Robert Hooke's Inverted Cone', *Proceedings: Mathematical, Physical and Engineering Sciences* Vol. **463**, No. 2081 (May 8, 2007), pp. 1259-1269 Published by: Royal Society
https://www.jstor.org/stable/20209181

22 Lambeck, K. (1988) *Geophysical Geodesy – The Slow deformations of the Earth*, Clarendon Press, Oxford,

23 Xiaoming Cui , Heping Sun , Jianqiao Xu, Jiangcun Zhou and Xiaodong Chen, Detection of free core nutation resonance variation in Earth tide from global superconducting gravimeter observations, *Earth, Planets and Space* (2018) 70:199 https://doi.org/10.1186/s40623-018-0971-9

24 Dunbavin, Paul (2005) *Under Ancient Skies*, Third Millennium Publishing, Nottingham, ISBN: 0-9525029-2-5 (see chapter 4)

This article was originally published as an interactive web-page in 2020 at:

https://www.third-millennium.co.uk/irishgodkings

ABOUT THE AUTHOR

Paul Dunbavin was born in 1954 in Derbyshire; educated in physics and computing. From 1974 to 1999 he pursued a career in computing and subsequently ran a business transfer agency in Yorkshire. He is broadly self-taught across the arts and sciences and was a former Mensa local-secretary in Aberdeen. His interest for over 35 years has been cross-disciplinary research into prehistory, which he has occasionally published in his books and various articles and papers. His work is well known among enthusiasts and academics and receives a mixture of both positive and negative reaction. He contributed to a History Channel television series called *Puzzles of the Past* in the 1990s and has also had several magazine articles published.

Being largely self-taught and broadly educated across the arts and sciences, he has always preferred to consider himself as a researcher first and an author second. In authorship his primary interests are astronomy, ancient history, mythology and catastrophism with a side interest in the ethnography of Scotland and the British Isles.

His first book *The Atlantis Researches* was published in 1995 and republished in second edition by Constable in 2002 as *Atlantis of the West*. This was followed by *Picts and Ancient Britons* in 1998 and *Under Ancient Skies* in 2005. He has also written a number of research papers and articles on related subjects. Although for some years out of physical print these books were made available again in 2017 in Kindle editions. A new book: *Towers of Atlantis* was published in 2017.

As an author, of non-fiction Paul Dunbavin prefers to write 'real books' rather than publishers' formula pap. He prefers to present evidence in an interesting way but with fully referenced source bibliography in the academic style. The reader can expect to find original theories and conclusions unique to the author within every chapter, yet all are entirely based upon source evidence and current standard textbook science. The author offers an alternative theory of catastrophism in prehistory that owes nothing to Velikovsky.

www.third-millennium.co.uk

www.ingramcontent.com/pod-product-compliance
Lightning Source LLC
Chambersburg PA
CBHW051048050726
47592CB00002B/434